Flow and Heat Transfer in Rotating-Disc Systems

Volume 1—Rotor-Stator Systems

MECHANICAL ENGINEERING RESEARCH STUDIES

ENGINEERING DESIGN SERIES

Series Editor: **Professor W. D. Morris,** *Department of Mechanical Engineering, University College, Swansea, Wales*

1. Flow and Heat Transfer in Rotating-Disc Systems
   Volume 1—Rotor-Stator Systems
   J. M. Owen *and* **R. H. Rogers**

Flow and Heat Transfer in Rotating-Disc Systems

Volume 1—Rotor-Stator Systems

J. M. Owen

and

R. H. Rogers

School of Engineering and Applied Sciences
University of Sussex, England

RESEARCH STUDIES PRESS LTD.
Taunton, Somerset, England

JOHN WILEY & SONS INC.
New York · Chichester · Toronto · Brisbane · Singapore

RESEARCH STUDIES PRESS LTD.
24 Belvedere Road, Taunton, Somerset, England TA1 1HD

Marketing and Distribution:

Australia, New Zealand, South-east Asia:
Jacaranda-Wiley Ltd., Jacaranda Press
JOHN WILEY & SONS INC.
GPO Box 859, Brisbane, Queensland 4001, Australia

Canada:
JOHN WILEY & SONS CANADA LIMITED
22 Worcester Road, Rexdale, Ontario, Canada

Europe, Africa:
JOHN WILEY & SONS LIMITED
Baffins Lane, Chichester, West Sussex, England

North and South America and the rest of the world:
JOHN WILEY & SONS INC.
605 Third Avenue, New York, NY 10158, USA

Library of Congress Cataloging-in-Publication Data

Owen, J. M.
Flow and heat transfer in rotating-disc systems / J. M. Owen and R. H. Rogers.
p. cm.—(Mechanical engineering research studies. Engineering design series; no. 1–)
Bibliography: p.
Includes index.
Contents: v. 1. Rotor-stator systems.
ISBN 0 471 92474 1 (Wiley)
1. Disks, Rotating. 2. Fluid dynamics. 3. Heat—Transmission. I. Rogers, Ruth H. II. Title. III. Series.
TJ1058.094 1989
620.1'064——dc20 89-10259
CIP

British Library Cataloguing in Publication Data

Owen, J. M. (John Michael, *1938–*)
Flow and heat transfer in rotating-disc systems.
Vol. 1, Rotor-stator systems
1. Rotating machinery. Fluid engineering & heat transfer
I. Title II. Rogers, R. H. (Ruth Helen, *1927–*) III. Series.
621.8'11

ISBN 0 86380 090 4

ISBN 0 86380 090 4 (Research Studies Press Ltd.)
ISBN 0 471 92474 1 (John Wiley & Sons Inc.)

Printed in Great Britain by Galliard (Printers) Ltd., Great Yarmouth

Contents

List of Figures

List of Tables

List of Commonly Used Symbols

The list below gives the (most common) meanings of those symbols that are used in many parts of the text. Occasionally a local meaning will be given: this will be made clear where it occurs in the text.

Subscripts

0	value on rotor in rotor-stator system or on stator in an infinite fluid
ad	value for adiabatic disc
av	radially-weighted average (except Nu_{av}, Nu^*_{av})
b,d,p,s	values for blade-cooling, disc-cooling, pre-swirl and sealing flows (Chapters 8, 9)
b	value at $r = b$
c	value in the core (except s_c, G_c)
$crit$	critical value
eff	effective value in turbulent flow
fd	free-disc value
I	value in incoming fluid
iso	value for isothermal distribution
lam	value in laminar flow
max, min	maximum, minimum value
$quad$	value for "purely quadratic" distribution
r,ϕ,z	radial, tangential, axial directions
ref	reference value, defined as appropriate
s	value on stator in rotor-stator system (except k_s, Re_s, δ_s)
$turb$	value in turbulent flow
w	value on disc under discussion (except C_w, K_w, Re_w)
∞	value far from disc

Superscript

$*$	(a) β^*; value of β with no superposed flow
	(b) Nu^*, Nu^*_{av}, Ec^*_b; values of Nu, Nu_{av}, Ec_b when dissipation is important

Constants

a	inner radius of disc
b	outer radius of disc
c_p	specific heat at constant pressure
k	thermal conductivity
k_s	effective height of roughness
s	axial gap between rotor and stator
s_c	clearance between shroud and rotor
μ	dynamic viscosity
ν	kinematic viscosity
ρ	density
Ω	angular speed of rotor

Variables

H	$c_pT + \frac{1}{2}(v_r^2 + v_\phi^2) + p/\rho$; total enthalpy (in the boundary layer of an incompressible fluid)
M	$-2\pi\int_a^b r^2 \tau_{\phi,0}\, dr$; moment on one side of the rotor
$\dot{m}$	radial mass flow-rate
p	static pressure in the fluid
q	heat flux in the axial direction
r	radial coordinate
T	temperature
(u, v, w)	velocity components referred to rotating cylindrical coordinates
(v_r, v_ϕ, v_z)	velocity components referred to stationary cylindrical coordinates
v^*	$(\tau_0/\rho)^{1/2}$; friction velocity
$\bar{W}$	average axial component of velocity of external flow
α	$-\tau_{r,0}/\tau_{\phi,0} = \lim_{z\to 0}[v_r(r,z)/v_\phi(r,z)]$
β, β^*	ω/Ω; core-swirl ratio

Γ	rv_ϕ; swirl
δ, δ_T	thickness of velocity, thermal boundary layer
ϵ_m, σ	"similarity parameters", $(\dot{m}/\mu r = \epsilon_m (x^2 Re_\phi)^\sigma)$
τ_0	shear stress at surface
(τ_r, τ_ϕ)	radial, tangential components of shear stress in boundary layer
ϕ	tangential coordinate
Φ	dissipation term in energy equation
ω	angular speed of fluid outside boundary layer

Nondimensional groups

C_M	$M/\frac{1}{2}\rho\Omega^2 b^5$; moment coefficient for one side of the rotor
$C_{p,max}$	$\Delta p_{max}/\frac{1}{2}\rho\bar{W}^2$, where Δp_{max} is the maximum circumferential pressure difference; nondimensional pressure asymmetry
C_w	$\dot{m}/\mu r$; nondimensional flow-rate
Ec	$\Omega^2 r^2/c_p \Delta T$, where ΔT is an appropriate temperature difference; Eckert number
Ec^*	$\Omega^2 r^2/\Delta H$, where ΔH is an appropriate enthalpy difference; modified Eckert number
G	s/b; gap ratio
G_c	s_c/b; shroud clearance ratio
K_H	$\int_0^\delta v_r (H - H_\infty)\,dz \Big/ (H_0 - H_\infty)\int_0^\delta v_r\,dz$; enthalpy shape factor
K_T	$\int_0^\delta v_r (T - T_\infty)\,dz \Big/ (T_0 - T_\infty)\int_0^\delta v_r\,dz$; temperature shape factor
K_v	$\int_0^\delta v_r (v_\phi - v_{\phi,\infty})\,dz \Big/ (v_{\phi,0} - v_{\phi,\infty})\int_0^\delta v_r\,dz$; velocity shape factor
K_w	$-\pi Re_\phi \dfrac{\tau_{\phi,0}}{\rho\Omega^2 b^2} \Big/ \dfrac{\dot{m}}{\mu r}$; stress-flow-rate parameter

Nu	$rq_w/k\Delta T$, where ΔT is an appropriate temperature difference; local Nusselt number
Nu^*	value of Nu when $\Delta T = T_0 - T_{0,ad}$
Nu_{av}	$bq_{w,av}/k\Delta T_{av}$; average Nusselt number
Nu^*_{av}	value of Nu^*_{av} when $\Delta T_{av} = (T_0 - T_{0,ad})_{av}$
Pr	$\mu c_p/k$; Prandtl number
R, R'	recovery factors
Re_s	$\rho\Omega s^2/\mu$; gap Reynolds number
Re_w	$\rho\bar{W}b/\mu$; external-flow Reynolds number
Re_ϕ	$\rho\Omega b^2/\mu$; rotational Reynolds number
Re'_ϕ	$\rho\omega b^2/\mu$; modified rotational Reynolds number
x	r/b; nondimensional radius
ζ	$z(\nu/\Omega)^{-1/2}$, z/δ or z/s; nondimensional axial coordinate
λ_{lam}	$C_w Re_\phi^{-1/2}$; laminar flow parameter
λ_{turb}	$C_w Re_\phi^{-4/5}$; turbulent flow parameter
χ, χ'	correction factors for heat flux

Note: the *local* rotational Reynolds number is not given a separate symbol, but is written everywhere as $x^2 Re_\phi$.

Author's Preface

Dorfman's (1963) book *Hydrodynamic resistance and heat loss of rotating solids* has became a standard reference for research engineers and designers involved with the fluid dynamics and heat transfer of rotating discs and cylinders. He assumed (as we do) that the reader would have an adequate understanding of "conventional" fluid mechanics and heat transfer. He included (as we hope we have) a reasonable mixture of mathematical arguments and experimental data enabling the reader to understand and to apply the subject.

Greenspan's (1968) monograph *The theory of rotating fluids* covered rotating flows in a more mathematical manner. Unlike Dorfman, whose equations were formulated for a stationary coordinate system, Greenspan used rotating coordinates to bring out the important concepts of linear equations (in which Coriolis accelerations dominate over the nonlinear inertia terms), Ekman layers and the Taylor-Proudman theorem. An oversimplification might be to say that Greenspan's book was more directed to the meteorologist and oceanographer: Dorfman's to the engineer.

In the past twenty years, numerous papers on rotating-disc systems have been published, but there has been no serious attempt to pull together the many ideas that have been developed. As we ourselves have contributed to the resulting "literature mountain", we thought it our responsibility to attempt to simplify, unify and extend the existing work and to produce a monograph that would provide a basic understanding of the subject and which could be used by research workers and engineers.

As well as being a fascinating subject of fundamental interest, flow in rotating-disc systems is also a topic of practical importance, not least in its application to flows inside turbomachinery. With the pursuit of higher thermal efficiencies for gas turbines, the increasing cycle temperatures have necessitated the use of air cooling for the hot rotating components. This has encouraged the gas-turbine companies to sponsor research programmes into effective cooling methods, and these have led to increasing interest in the flow inside rotating-disc systems. (Ironically, it is the increase in our involvement with rotating-disc research over the past few years that has prevented us from completing this monograph within the time limit that was originally set!)

Complementary to the commercial pressures that determine the directions of research, improvement in instrumentation (particularly data-loggers and nonintrusive laser-doppler anemometers) has enabled large quantities of accurate experimental data to be obtained. The increase in computing power has also made it possible to obtain numerical solutions of the parabolic boundary-layer equations and the elliptic Navier-Stokes equations for an increasing number of rotating flows. However, we observe wryly that the old-fashioned and unfashionable momentum-integral equations continue to provide solutions that are often more useful than those obtained from the modern elliptic solvers!

As most of our research on rotating-disc systems has been sponsored by gas-turbine companies, it is inevitable that the choice of topics in this monograph is somewhat biased. However, we believe that much of the material can be applied to rotating flows far removed from the gas turbine. In fact, in the second volume, we have taken ideas that were originally developed for problems arising in meteorology and oceanography and have applied them to the gas turbine: we hope that the converse might also be possible!

We should like to take this opportunity to thank those companies (particularly Rolls Royce plc and Ruston Gas Turbines plc) who have presented us with many intriguing problems and who also, together with the Science and Engineering Research Council, have supplied us with the money to investigate those problems. In addition, Rolls Royce kindly gave us their permission to present Figures 1.1 and 1.2 in our monograph.

We are also indebted to our colleagues (academic and technical) and research students, past and present, who have contributed a great deal of the material used in this monograph and from whom we have learned so much. In particular, we wish to thank Dr C L Ong for helping to prepare some of the figures and Mrs P Cherry for converting rough drawings into finished tracings. Last, but by no means least, one author wishes to thank the other for having the patience to produce the entire camera-ready copy!

J M Owen
R H Rogers
April 1989

CHAPTER 1
Introduction

The main objects of this monograph are to give an understanding of some of the rotating flows that occur inside turbomachinery and to provide ideas and information for research workers and design engineers.

Important examples of the application of rotating flows can be found in the internal air-systems of gas-turbine engines, simplified diagrams of which are shown in Figures 1.1 and 1.2. Since the development of these engines in the 1940s, turbine-entry temperatures have increased from around 900 K to over 1600 K. This increase has improved the thermal efficiency, resulting in improved performance and reduced fuel consumption. Whilst some of the increase in temperature has been made possible through the development of new materials, most has occurred because of improvements in cooling technology. For this purpose, a small percentage of air, extracted from the compressor, is used to cool the nozzle guide vanes, turbine blades and the discs to which the blades are attached.

As the turbine-entry temperatures have increased, so have the compression ratios; this has resulted in an increase in the temperature of the air available for cooling purposes. In fact, the temperature of the air leaving the compressor of a modern engine is higher than the turbine-entry temperatures of the earliest engines!

Thus as effective cooling has become more necessary, it has also become more difficult and expensive to achieve.

Most of the research into air-cooled gas turbines has concentrated on the nozzles and blades, but over the past few years attention has turned to the turbine discs. Further fuel savings can be made by reducing the amount of (expensive) compressed air that is used for cooling and sealing these discs. Referring to Figure 1.2, it can be seen that a turbine disc usually rotates next to either a stationary casing or another rotating disc. It is the flow and heat transfer between a rotating and a stationary disc that is the subject of Volume I of this monograph; the flow inside the cavity between two corotating discs is dealt with in Volume II.

Typical problems that concern the engineer are the stress levels, fatigue life and radial growth of the turbine discs. Prediction of these requires knowledge of the temperature distribution inside the discs for a range of operating conditions, and the effective thermal

---

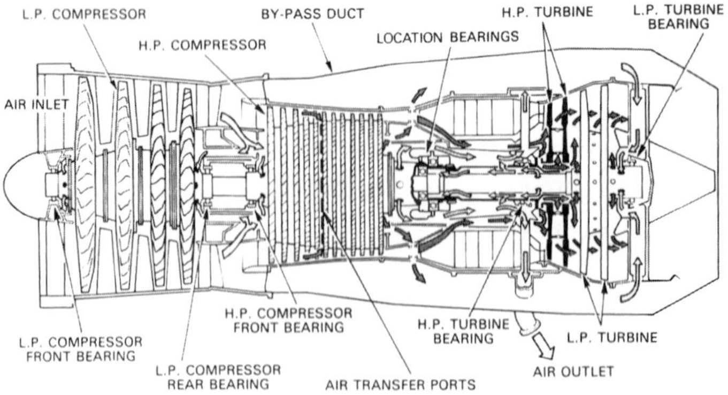

Figure 1.1 Internal air system in a gas-turbine engine

Arrows: flow of cooling air extracted from compressor

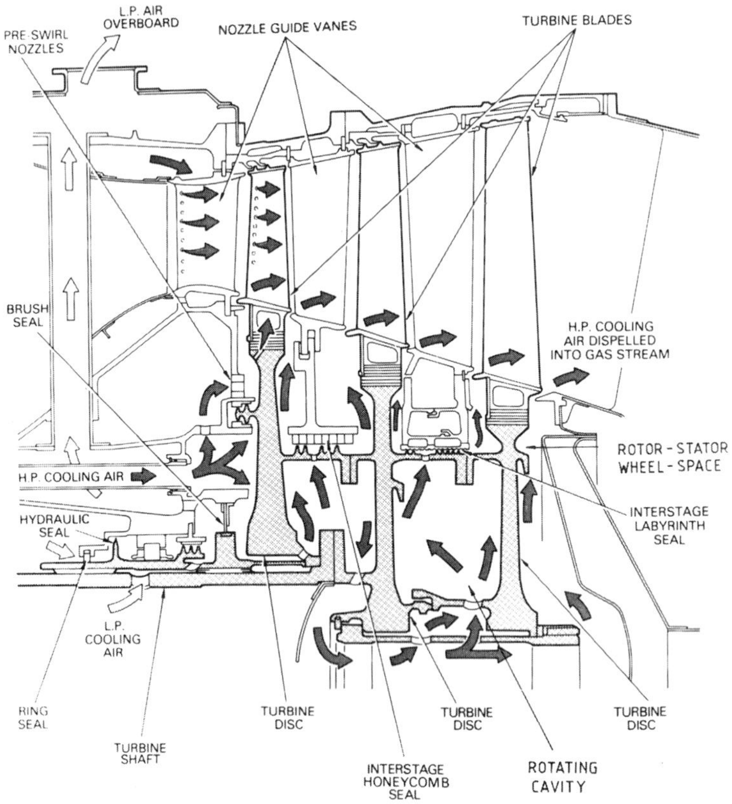

Figure 1.2 Schematic diagram of a turbine cooling and sealing system

Arrows: flow of low pressure (L.P.) and high pressure (H.P.) cooling air extracted from compressor

---

boundary conditions can be determined accurately only if the fluid dynamics are understood.

The external mainstream gas which flows over the turbine blades may be at a temperature well above the melting point of the nickel alloys from which the blades are made. The ingestion of this gas, through the clearances between the rotating and stationary seals, into the wheel-space between the turbine disc and its adjacent casing (see Figure 1.2) can cause overheating with a consequent reduction in disc life and, under extreme conditions, the loss of turbine blades. This ingestion can be reduced by using a radial outflow of disc-cooling air to pressurize the wheel-space. In order to minimize the amount of air necessary to seal the system, the designer needs to know how the sealing flow-rate depends on the system geometry, on the seal clearances, on the rotational speed of the disc and on the conditions in the external mainstream.

As shown in Figure 1.2, the cooling air fed to the turbine blades is often supplied through pre-swirl nozzles in the stationary casing adjacent to the rotating disc. The nozzles impart a tangential component of velocity to the air so that it rotates at a speed close to that of the disc, thereby reducing the temperature of the air relative to the rotating blades. However, in passing from the nozzles to the blade-cooling passages in the disc, the pre-swirl air may be "contaminated" with the disc coolant or with ingested mainstream gas. The design engineer needs to minimize this contamination, and he needs to be able to calculate the effective temperature of the air reaching the blades.

These are just some examples of the difficulties that beset the designer of internal air-systems. However, before solving these problems for the complex geometries of an actual engine, it is useful to study the flows that occur in simple plane-disc systems. By understanding the basic flows, and by using approximate mathematical models, it is possible to improve the design of air-systems and to become less dependent on empiricism and on the extrapolation of existing designs. Similarly, it should be possible to apply some of the methods and information in this monograph to a variety of problems that have been neither considered nor envisaged by the authors!

Chapter 2 provides the basic equations that are used throughout the book. The elliptic Navier-Stokes and energy equations (and their turbulent equivalents) are reduced to a differential boundary-layer form, and are then converted to integral equations suitable for rotating-disc systems. Use is also made of the so-called linear equations (which are employed extensively in geophysical systems) and of the Reynolds analogy between the transfer of heat and momentum.

Chapters 3, 4 and 5 deal with flow and heat transfer for a single disc. (Some of the numerical techniques used in these chapters are relegated to the appendices.) The simplest case is the so-called *free disc*: a disc rotating in a quiescent fluid. This problem, which was first addressed by von Kårmån (1921), provides one of the few examples for which exact solutions of the Navier-Stokes equations exist. The problem can be readily extended to the case of a single disc (either rotating or stationary) in a rotating fluid, and this in turn can be applied to the flow between a rotating and a stationary disc.

Chapters 6, 7, 8 and 9 concentrate on flow and heat transfer in rotor-stator systems with or without a superposed flow of fluid. The particular geometries and topics discussed in these chapters largely reflect the authors' interests and depend on the information available in the open literature.

There are some rotor-stator systems about which little is known: local heat-transfer measurements in systems with a radial inflow of fluid are poorly represented; for the inflow case, even the flow structure has not been fully investigated. The "ingress problem", in which there is ingestion of external fluid into the rotor-stator wheel-space, is a topic for which there are many questions but only a few answers. It is appropriate, therefore, that this book should end with the ingress problem: a topic that is currently being investigated by many research workers throughout the world.

CHAPTER 2
Basic Equations

The Navier-Stokes equations and the energy equation provide the basis from which all newtonian flow can be determined. The boundary-layer approximations are often valid near a solid surface, and the resulting equations may be expressed in either differential or partially integrated form: the integral equations provide a relatively simple description of the flow, and are of particular use when the flow is turbulent. For flows involving heat transfer, the Reynolds analogy can also produce simplification and may, under some conditions, obviate the need to solve the energy equation. In this chapter a summary of these equations and techniques is given, with special reference to their application to flows in which rotation is an important factor.

At the surface of a rotating disc, it is the convention in this book to use a subscript 0; at the surface of the stationary disc of a rotor-stator system, the subscript s is used. However, throughout most of this chapter (and also throughout Chapters 3, 4 and 5, where flow near a single disc is discussed), it is convenient to use the subscript 0 for every case.

The main purpose of this chapter is to give a basis of reference for the later discussion of "real-life" problems. While the basic equations and approximations are discussed only briefly, the methods of their solution (which occur frequently in later parts of the book) are explained in some detail. This will enable readers to study these methods if they wish, or to take them on trust if they prefer to do so.

2.1 DIFFERENTIAL EQUATIONS FOR AXISYMMETRIC FLOW

For the flow of the fluid between the discs in a rotor-stator system it is usually convenient to use a stationary frame of reference. As shown in Figure 2.1, the

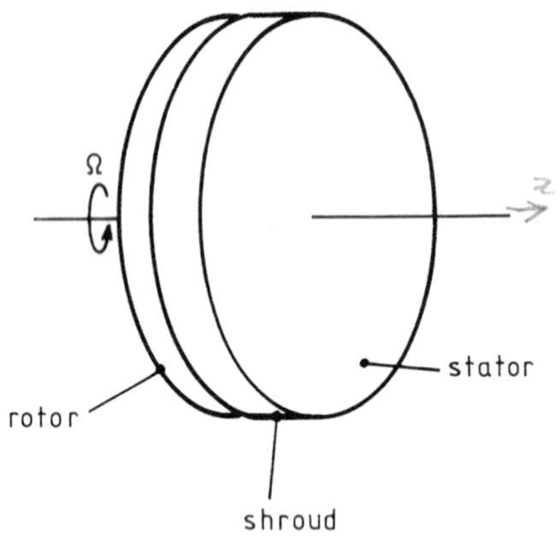

Figure 2.1 Schematic diagram of a rotor-stator system (shroud attached to stator)

rotor rotates about the z-axis and the origin O is taken as the point where the axis of rotation intersects the rotor. The direction of the z-axis is chosen to point towards the stator: thus the rotor lies in the plane $z = 0$ and the stator in the plane $z = s$, where s is the distance between the two discs; each disc is of radius b. In the conventional fashion, distances from Oz are represented by the variable r, and the cylindrical polar coordinates (r, ϕ, z) are such that ϕ increases in the direction of a right-handed screw relative to the positive direction of Oz. For convenience, the rotor is assumed to be rotating in the direction of ϕ increasing, its angular speed being Ω (radians/sec). Referred to the fixed frame of reference, the velocity of the fluid at the point (r,ϕ,z) is taken to be $\mathbf{v} = (v_r, v_\phi, v_z)$, its density ρ, its temperature T and its pressure p. (The velocity at a point on the rotor, at a distance r from the axis, is $\mathbf{v} = (0,\Omega r,0)$.) The discussion will be limited to the steady axisymmetric case: that is $\partial f/\partial t = \partial f/\partial \phi = 0$ for all dependent variables f.

There is often a cylindrical shroud at $r = b$ $(0 < z < s)$. This shroud may be attached to the stator (where its velocity is zero), or it may be attached to the rotor (in which case its velocity is $(0,\Omega b,0)$). Another possibility is for the shroud to be split into two parts, one attached to the stator and the other to the rotor.

2.1.1 Equations of motion: Navier-Stokes equations; Reynolds equations

It will be assumed that the flow is described by the continuity and Navier-Stokes equations. Only the equations for incompressible flow will be presented here, and the dynamic viscosity μ and kinematic viscosity ν will be assumed to be constant. Then the equations may be written (see Schlichting 1979)

$$\operatorname{div} \mathbf{v} = 0, \tag{2.01}$$

$$(\mathbf{v}.\nabla)\mathbf{v} = -\frac{1}{\rho}\operatorname{grad} p + \nu \nabla^2 \mathbf{v}. \tag{2.02}$$

In component form, these equations become

$$\frac{\partial v_r}{\partial r} + \frac{v_r}{r} + \frac{\partial v_z}{\partial z} = 0, \tag{2.03}$$

$$v_r \frac{\partial v_r}{\partial r} + v_z \frac{\partial v_r}{\partial z} - \frac{v_\varphi^2}{r} = -\frac{1}{\rho}\frac{\partial p}{\partial r} + \nu\left(\nabla^2 v_r - \frac{v_r}{r^2}\right), \qquad (2.04)$$

$$v_r \frac{\partial v_\varphi}{\partial r} + v_z \frac{\partial v_\varphi}{\partial z} + \frac{v_r v_z}{r} = \nu\left(\nabla^2 v_\varphi - \frac{v_\varphi}{r^2}\right), \qquad (2.05)$$

$$v_r \frac{\partial v_z}{\partial r} + v_z \frac{\partial v_z}{\partial z} = -\frac{1}{\rho}\frac{\partial p}{\partial z} + \nu \nabla^2 v_z , \qquad (2.06)$$

where

$$\nabla^2 = \frac{\partial^2}{\partial r^2} + \frac{1}{r}\frac{\partial}{\partial r} + \frac{\partial^2}{\partial z^2}. \qquad (2.07)$$

When the flow is laminar, these equations are appropriate as they stand. For turbulent flow, it is necessary to modify them to take account of the fact that the velocity and pressure each consists of two parts: there is an "average" component, and also a fluctuating component whose average (over some suitable time interval t_0, which is small compared with Ω^{-1}) is zero. If the fluctuating parts are indicated by the use of primes, the variables $\mathbf{v}$ and p in equations (2.01) and (2.02) are replaced by $\bar{\mathbf{v}} + \mathbf{v}'$ and $\bar{p} + p'$ respectively. The equations are then averaged (over the time interval t_0) and, assuming that the average values of the derivatives of $\mathbf{v}'$ and p' are also zero, they become

$$\operatorname{div} \bar{\mathbf{v}} = 0, \qquad (2.08)$$

$$(\bar{\mathbf{v}}.\nabla)\bar{\mathbf{v}} = -\frac{1}{\rho}\operatorname{grad}\bar{p} + \nu \nabla^2 \bar{\mathbf{v}} - \overline{(\mathbf{v}'.\nabla)\mathbf{v}'}, \qquad (2.09)$$

where the overbars denote average values.

The extra term $\overline{(\mathbf{v}'.\nabla)\mathbf{v}'}$ on the right-hand side of equation (2.09) is equivalent to an additional stress (in addition to the stresses due to the mean pressure and the viscous forces) which affects the average motion in turbulent flow. It is usually referred to as the Reynolds stress (and its components as the Reynolds stresses). In order to deal with the extra term, assumptions need to be made about its relationship with the mean velocity $\overline{\mathbf{v}}$. There is no known method of doing this which works in all situations, and empirical considerations have to be taken into account. In this monograph, the relevant relationships are given in the chapters where they occur.

In component form, equations (2.08) and (2.09) are

$$\frac{\partial \overline{v}_r}{\partial r} + \frac{\overline{v}_r}{r} + \frac{\partial \overline{v}_z}{\partial z} = 0, \tag{2.10}$$

$$\begin{aligned}\overline{v}_r \frac{\partial \overline{v}_r}{\partial r} + \overline{v}_z \frac{\partial \overline{v}_r}{\partial z} - \frac{\overline{v}_\phi^2}{r} &= -\frac{1}{\rho}\frac{\partial \overline{p}}{\partial r} + \nu\left(\nabla^2 \overline{v}_r - \frac{\overline{v}_r}{r^2}\right)\\ &\quad - \frac{1}{r}\frac{\partial}{\partial r}\left(r\overline{v_r'^2}\right) - \frac{\partial}{\partial z}\left(\overline{v_r' v_z'}\right) + \frac{\overline{v_\phi'^2}}{r},\end{aligned} \tag{2.11}$$

$$\begin{aligned}\overline{v}_r \frac{\partial \overline{v}_\phi}{\partial r} + \overline{v}_z \frac{\partial \overline{v}_\phi}{\partial z} + \frac{\overline{v}_r \overline{v}_\phi}{r} &= \nu\left(\nabla^2 \overline{v}_\phi - \frac{\overline{v}_\phi}{r^2}\right)\\ &\quad - \frac{1}{r}\frac{\partial}{\partial r}\left(r\overline{v_\phi' v_r'}\right) - \frac{\partial}{\partial z}\left(\overline{v_\phi' v_z'}\right) - \frac{\overline{v_r' v_\phi'}}{r},\end{aligned} \tag{2.12}$$

$$\begin{aligned}\overline{v}_r \frac{\partial \overline{v}_z}{\partial r} + \overline{v}_z \frac{\partial \overline{v}_z}{\partial z} &= -\frac{1}{\rho}\frac{\partial \overline{p}}{\partial z} + \nu \nabla^2 \overline{v}_z\\ &\quad - \frac{1}{r}\frac{\partial}{\partial r}\left(r\overline{v_r' v_z'}\right) - \frac{\partial}{\partial z}\left(\overline{v_z'^2}\right).\end{aligned} \tag{2.13}$$

2.1.2 Equations of motion referred to a rotating frame of reference: geostrophic approximation

It is sometimes convenient to use a frame of reference rotating with angular speed Ω' (Ω' may be the same as Ω, but not necessarily) about the z-axis; the velocity components referred to such a frame are written as (u, v, w); then it is clear that

$$u = v_r, \quad v = v_\phi - \Omega' r, \quad w = v_z, \tag{2.14}$$

and that the pressure p is unchanged. Substituting in equations (2.03) to (2.06), the equations of motion become

$$\frac{\partial u}{\partial r} + \frac{u}{r} + \frac{\partial w}{\partial z} = 0, \tag{2.15}$$

$$u\frac{\partial u}{\partial r} + w\frac{\partial u}{\partial z} - \frac{v^2}{r} - 2\Omega' v - \Omega'^2 r = -\frac{1}{\rho}\frac{\partial p}{\partial r} + \nu\left(\nabla^2 u - \frac{u}{r^2}\right), \tag{2.16}$$

$$u\frac{\partial v}{\partial r} + w\frac{\partial v}{\partial z} + \frac{uv}{r} + 2\Omega' u = \nu\left(\nabla^2 v - \frac{v}{r^2}\right), \tag{2.17}$$

$$u\frac{\partial w}{\partial r} + w\frac{\partial w}{\partial z} = -\frac{1}{\rho}\frac{\partial p}{\partial z} + \nu\nabla^2 w. \tag{2.18}$$

The terms $-2\Omega' v$, $2\Omega' u$ in equations (2.16), (2.17) respectively are usually referred to as the *Coriolis terms*. The equations are useful when the velocity components u, v, w and their derivatives are small compared with $\Omega' r$ (that is, when the nonlinear inertial accelerations are small compared with the Coriolis terms). In this case, all the nonlinear terms in equations (2.16) to (2.18) may be neglected and the equations written in the form

$$-2\Omega' v = -\frac{1}{\rho}\frac{\partial}{\partial r}(p - \tfrac{1}{2}\Omega'^2 r^2) + \nu\left(\nabla^2 u - \frac{u}{r^2}\right), \tag{2.19}$$

$$2\Omega' u = \nu\left(\nabla^2 v - \frac{v}{r^2}\right) \tag{2.20}$$

and

$$0 = -\frac{1}{\rho}\frac{\partial p}{\partial z} + \nu\nabla^2 w. \tag{2.21}$$

When viscous forces are negligible, this is referred to as the geostrophic approximation (especially in meteorological and oceanographical problems). In this case, equation (2.15) together with equations (2.19) to (2.21) give the interesting result that

$$u = 0, \quad \frac{\partial v}{\partial z} = 0, \quad \frac{\partial w}{\partial z} = 0. \tag{2.22}$$

This is a special case of the so-called Taylor-Proudman

theorem (see, for example, Batchelor 1967 or Greenspan 1968) which states that, for small relative velocities of a rotating inviscid fluid, all components of velocity are independent of z.

2.1.3 Energy equation for laminar and turbulent flow

The assumption of incompressibility made in the last section means that the density ρ may be regarded as constant throughout the fluid. Clearly this is not exactly true but, when variations in pressure and temperature are not too great, the assumption gives a good approximation to the real situation. It is also assumed that other thermodynamic properties of the fluid (the viscosity μ, the thermal conductivity k and the specific heat at constant pressure c_p) are constant.

Even if small temperature variations cause no significant change in density, they may result in the flow of heat from one part of the fluid to another. This may occur through either convection or conduction; there will be an additional term due to viscous dissipation. For laminar flow, the temperature T satisfies the equation

$$v_r \frac{\partial T}{\partial r} + v_z \frac{\partial T}{\partial z} = \frac{k}{\rho c_p} \nabla^2 T + \frac{1}{\rho c_p} \mu \Phi, \tag{2.23}$$

where

$$\Phi = 2\left(\frac{\partial v_r}{\partial r}\right)^2 + 2\left(\frac{v_r}{r}\right)^2 + 2\left(\frac{\partial v_z}{\partial z}\right)^2 + \left(\frac{\partial v_\phi}{\partial r} - \frac{v_\phi}{r}\right)^2 + \left(\frac{\partial v_\phi}{\partial z}\right)^2 + \left(\frac{\partial v_z}{\partial r} + \frac{\partial v_r}{\partial z}\right)^2; \tag{2.24}$$

these equations are given (with one typographical error!) by Dorfman (1963). The left-hand side of equation (2.23) corresponds to the convection of heat, the first term on the right-hand side to conduction, and Φ is the viscous-dissipation term; the latter is often (but not always) so small that it may be neglected.

For turbulent flow, the average temperature $\bar{T}$ satisfies the equation

$$\bar{v}_r \frac{\partial \bar{T}}{\partial r} + \bar{v}_z \frac{\partial \bar{T}}{\partial z} = \frac{k}{\rho c_p} \nabla^2 \bar{T} - \frac{1}{r}\frac{\partial}{\partial r}(r\overline{v_r' T'}) - \frac{\partial}{\partial z}(\overline{v_z' T'}) + \frac{1}{\rho c_p}\mu\bar{\Phi}, \tag{2.25}$$

where the dissipation due to the fluctuating part of the turbulent velocity is included in $\bar{\Phi}$. The extra terms on the right-hand side of equation (2.25) give the contribution to the heat flow caused by the fluctuating part of the turbulent velocity and temperature.

2.1.4 Inviscid equations

For a single disc in a fluid of large extent, or for a rotor-stator system whose clearance is sufficiently large, viscous and turbulent effects are negligible except in thin boundary layers near the discs. It is useful, therefore, to consider the solutions of the equations of motion when viscous terms are negligible: the subscript c is used to indicate that the flow considered is in the inviscid *core* of a rotor-stator system. When a single disc is discussed (as in this chapter and in Chapters 3, 4 and 5), the subscript ∞ is used for the flow far from the disc.

Equation (2.03) is unchanged, and equations (2.04) to (2.06) become

$$v_{r,\infty}\frac{\partial v_{r,\infty}}{\partial r} + v_{z,\infty}\frac{\partial v_{r,\infty}}{\partial z} - \frac{v_{\phi,\infty}^2}{r} = -\frac{1}{\rho}\frac{\partial p_\infty}{\partial r}, \tag{2.26}$$

$$v_{r,\infty}\frac{\partial v_{\phi,\infty}}{\partial r} + v_{z,\infty}\frac{\partial v_{\phi,\infty}}{\partial z} + \frac{v_{r,\infty}v_{\phi,\infty}}{r} = 0, \tag{2.27}$$

$$v_{r,\infty}\frac{\partial v_{z,\infty}}{\partial r} + v_{z,\infty}\frac{\partial v_{z,\infty}}{\partial z} = -\frac{1}{\rho}\frac{\partial p_\infty}{\partial z}. \tag{2.28}$$

In practice, the radial component of velocity is zero throughout much of the core; in this case, it follows from equation (2.03) that $v_{z,\infty}$ is independent of z and from equation (2.27) that $v_{\phi,\infty}$ is independent of z. The Taylor-Proudman result, which is a consequence of the geostrophic approximation (as described in Section 2.1.2), is therefore valid when the full equations are used in the core, provided that $v_{r,\infty} = 0$. Equation (2.28) shows that p_{∞} is independent of z and equation (2.26) gives a relationship between the radial pressure gradient and $v_{\phi,\infty}$. These results are summarized as

$$v_{r,\infty} = 0, \quad \frac{\partial v_{\phi,\infty}}{\partial z} = 0, \quad \frac{\partial v_{z,\infty}}{\partial z} = 0, \quad \frac{1}{\rho}\frac{\partial p_{\infty}}{\partial r} = \frac{v_{\phi,\infty}^2}{r}. \qquad (2.29)$$

When there is a superposed radial flow in a rotor-stator system, it often happens that there is a region of the core in which there is no radial component of velocity, because all the incoming fluid has been entrained into the boundary layer on the rotor. However, near the inlet, before all the superposed fluid has been entrained, there may be a "source region" in which there is a radial component of velocity within the inviscid core. Although various models can be considered for the flow in the source region, the authors know of no adequate theory for the general case. Such a theory would have to take account of the possibility of recirculating fluid as well as the fact that the boundary layer on the stator usually emits fluid into the core.

Although not used later in this book, the system of equations derived below may be of use in certain cases for dealing with the source region. They are based on the *assumptions* that the angular momentum of the incoming fluid is conserved and that the tangential component of velocity is independent of z: this means that a *free vortex* occurs in which

$$rv_{\phi,\infty} = K, \tag{2.30}$$

where K is independent of r and z. A reduced pressure, P can be defined as

$$P = p_{\infty} - \frac{1}{2}\rho v_{\phi,\infty}^2 = p_{\infty} - \frac{1}{2}\rho\left(\frac{K}{r}\right)^2; \tag{2.31}$$

then equations (2.03), (2.26) and (2.28) give

$$\frac{\partial v_{r,\infty}}{\partial r} + \frac{v_{r,\infty}}{r} + \frac{\partial v_{z,\infty}}{\partial z} = 0, \tag{2.32}$$

$$v_{r,\infty}\frac{\partial v_{r,\infty}}{\partial r} + v_{z,\infty}\frac{\partial v_{r,\infty}}{\partial z} = -\frac{\partial P}{\partial r}, \tag{2.33}$$

$$v_{r,\infty}\frac{\partial v_{z,\infty}}{\partial r} + v_{z,\infty}\frac{\partial v_{z,\infty}}{\partial z} = -\frac{\partial P}{\partial z}; \tag{2.34}$$

equation (2.27) is automatically satisfied by the value of $v_{\phi,\infty}$ defined in equation (2.30). These equations are similar to, but not identical with, those for plane two-dimensional flow with p replaced by the reduced pressure P. Equations (2.32) to (2.34) can, in principle, be solved independently of the value of K which occurs only implicitly in the pressure term.

2.2 BOUNDARY-LAYER EQUATIONS

2.2.1 Equations in a stationary frame

If the distance between the discs is small compared with their radius, it may be assumed that

(i) the component of velocity v_z is very much smaller in magnitude than either of the other two components;

(ii) the rate of change of any variable (other than the pressure) in the direction normal to the disc is much greater than its rate of change in a radial or in a tangential direction;

(iii) the pressure depends only on distance from the axis of rotation.

Similar approximations are valid when the flow between the discs is such that viscous effects are negligible except in thin layers close to the discs - that is, in boundary layers. Since this situation arises in a number of cases, it is worth discussing the effect of these approximations on the equations of Section 2.1. It should be noted, in passing, that the boundary-layer equations for the discs will not be valid in the neighbourhood of the shroud nor, very often, near the axis of rotation.

The effect of retaining only derivatives with respect to z (compared with those with respect to r and ϕ) is to reduce the viscous and Reynolds stresses to a vector τ, the *shear stress*, whose components are $(\tau_r, \tau_\phi, \tau_z)$. Similarly the heat flux becomes a scalar quantity, q, say. The approximate equations for both laminar and turbulent flow are conveniently collapsed into a single set by omitting the overbars on the average velocity components, pressure and temperature; in addition it should be noted that the fluctuating terms in the expressions for τ and q are identically zero in laminar flow.

Equations (2.03) and (2.10) are unchanged by the approximation, but are repeated here, for completeness, as equation (2.35). The two sets of equations (2.04) to (2.06) and (2.11) to (2.13) reduce to equations (2.36) to (2.38), and equations (2.23) and (2.24) become equation (2.39).

$$\frac{\partial v_r}{\partial r} + \frac{v_r}{r} + \frac{\partial v_z}{\partial z} = 0, \qquad (2.35)$$

$$v_r \frac{\partial v_r}{\partial r} + v_z \frac{\partial v_r}{\partial z} - \frac{v_\phi^2}{r} = -\frac{1}{\rho}\frac{\partial p}{\partial r} + \frac{1}{\rho}\frac{\partial \tau_r}{\partial z}, \qquad (2.36)$$

$$v_r \frac{\partial v_\phi}{\partial r} + v_z \frac{\partial v_\phi}{\partial z} + \frac{v_r v_\phi}{r} = \frac{1}{\rho}\frac{\partial \tau_\phi}{\partial z}, \qquad (2.37)$$

$$0 = -\frac{\partial p}{\partial z}, \qquad (2.38)$$

$$v_r \frac{\partial T}{\partial r} + v_z \frac{\partial T}{\partial z} = -\frac{1}{\rho c_p}\frac{\partial q}{\partial z} + \frac{1}{\rho c_p}\mu\Phi, \qquad (2.39)$$

where

$$\tau_r = \mu \frac{\partial v_r}{\partial z} - \rho \overline{v_r' v_z'}, \tag{2.40}$$

$$\tau_\phi = \mu \frac{\partial v_\phi}{\partial z} - \rho \overline{v_\phi' v_z'}, \tag{2.41}$$

$$q = -\left(k \frac{\partial T}{\partial z} - \rho c_p \overline{T' v_z'}\right), \tag{2.42}$$

$$\mu\Phi = \tau_r \frac{\partial v_r}{\partial z} + \tau_\phi \frac{\partial v_\phi}{\partial z}. \tag{2.43}$$

Since one of the approximations of boundary-layer theory is that the pressure p is independent of z throughout a boundary layer, it follows that, for a single disc, $p = p_\infty$ within the layer, p_∞ being the value of the pressure far from the disc; for the discs of a rotor-stator system, p_∞ is replaced by p_c, the pressure in the inviscid core. Using equation (2.29) it can be seen that, when the flow far from a single disc (or in the core for a two-disc system) is inviscid, the term involving $\partial p/\partial r$ on the right-hand side of equation (2.36) can be replaced by $-v_{\phi,\infty}^2/r$ (or $-v_{\phi,c}^2/r$).

Using these results, together with the equation of continuity (2.35), it is often convenient to write equations (2.35) to (2.37) in the form

$$\frac{\partial}{\partial r}(r v_r) + \frac{\partial}{\partial z}(r v_z) = 0, \tag{2.44}$$

$$\frac{\partial}{\partial r}(r v_r^2) + \frac{\partial}{\partial z}(r v_r v_z) - v_\phi^2 = - v_{\phi,\infty}^2 + \frac{r}{\rho}\frac{\partial \tau_r}{\partial z}, \tag{2.45}$$

$$\frac{\partial}{\partial r}(r^2 v_r v_\phi) + \frac{\partial}{\partial z}(r^2 v_z v_\phi) = \frac{r^2}{\rho}\frac{\partial \tau_\phi}{\partial z}, \tag{2.46}$$

and equation (2.39) as

$$\frac{\partial}{\partial r}(r v_r T) + \frac{\partial}{\partial z}(r v_z T) = - \frac{r}{\rho c_p}\frac{\partial q}{\partial z} + \frac{r}{\rho c_p}\mu\Phi. \tag{2.47}$$

It is sometimes convenient to replace the temperature T in equation (2.47) by the *total enthalpy*, H, where, for an incompressible fluid,

$$H = c_p T + \tfrac{1}{2}(v_r^2 + v_\phi^2) + \frac{p}{\rho}. \qquad (2.48)$$

It should be noted that this definition of total enthalpy is valid only in the boundary layer when $|v_z| \ll (v_r^2 + v_\phi^2)^{\frac{1}{2}}$: this is true when there is no suction or blowing through the disc. (For a perfect gas, the pressure term in the definition of H is not present: see Schlichting 1979 for a fuller discussion of this.) Equation (2.47) then becomes

$$\frac{\partial}{\partial r}(rv_r H) + \frac{\partial}{\partial z}(rv_z H) = -\frac{r}{\rho}\frac{\partial}{\partial z}(q - \tau_r v_r - \tau_\phi v_\phi). \qquad (2.49)$$

For laminar flow, the expressions for τ_r, τ_ϕ and q in equations (2.40) to (2.42) contain no averaged fluctuating terms; they may be used, without further discussion, in equations (2.44) to (2.47) to give a self-consistent set of equations. For turbulent flow, however, progress cannot be made without first making some assumptions about the averaged fluctuating terms. A common set of assumptions (see Schlichting 1979) involves writing

$$\rho\overline{v_r' v_z'} = -\epsilon_r \frac{\partial v_r}{\partial z}, \quad \rho\overline{v_\phi' v_z'} = -\epsilon_\phi \frac{\partial v_\phi}{\partial z}, \quad \rho\overline{T' v_z'} = -\epsilon_T \frac{\partial T}{\partial z}. \qquad (2.50)$$

Then equations (2.40) to (2.42) become

$$\tau_r = \mu_{r,\mathrm{eff}} \frac{\partial v_r}{\partial z}, \quad \tau_\phi = \mu_{\phi,\mathrm{eff}} \frac{\partial v_\phi}{\partial z}, \quad q = -k_{\mathrm{eff}} \frac{\partial T}{\partial z}, \qquad (2.51)$$

where

$$\mu_{r,\mathrm{eff}} = \mu + \epsilon_r, \quad \mu_{\phi,\mathrm{eff}} = \mu + \epsilon_\phi, \quad k_{\mathrm{eff}} = k + c_p \epsilon_T. \qquad (2.52)$$

The coefficients $\mu_{r,\mathrm{eff}}$ and $\mu_{\phi,\mathrm{eff}}$ are "effective" viscosities, and k_{eff} is an "effective" conductivity. They give formal simplification to the turbulent case. However, the problem has merely been shifted because the values of ϵ_r, ϵ_ϕ and ϵ_T are not known: indeed they are not, in general, constants. However, for the special case of isotropic turbulence ($\epsilon_r = \epsilon_\phi = \epsilon$, say), it is common to define a *turbulent Prandtl number*, $Pr_{turb} = \epsilon/\epsilon_T$, by analogy with the laminar Prandtl number (see list of symbols); Pr_{turb} is often assumed to be constant.

The simplest turbulence model, originally proposed by Prandtl (1921), is that the coefficients ϵ are proportional to z: this follows from an elementary mixing-length hypothesis analogous to that used in molecular theory. More sophisticated versions of the mixing-length hypothesis have been developed, as have alternatives to such a theory, and these are discussed elsewhere.

2.2.2 Ekman-layer equations

When equations (2.19) to (2.21), referred to a frame of reference rotating with angular speed Ω', are valid, the boundary-layer equations simplify to

$$-2\Omega'(v - v_\infty) = \frac{1}{\rho}\frac{\partial\tau_r}{\partial z}, \tag{2.53}$$

$$2\Omega' u = \frac{1}{\rho}\frac{\partial\tau_\phi}{\partial z} \tag{2.54}$$

where τ_r, τ_ϕ are given by equations (2.40) and (2.41), with v_r, v_ϕ replaced by u, $v + \Omega r$ respectively. These equations were first derived by Ekman (1905) in connection with problems of flow in the ocean and, as will be seen in later chapters, have a particularly simple analytic solution for laminar flow.

By integrating equation (2.54) across the boundary layer, a useful relationship can easily be obtained between the tangential component of shear stress at the disc, $\tau_{\phi,0}$, and the local mass flow-rate, $\dot{m}_0$, through the boundary layer, where

$$\dot{m}_0 = 2\pi\rho r\int_0^\infty u\,dz. \tag{2.55}$$

(It is assumed here that there is no radial component of velocity outside the boundary layer.) This relationship may be written as

$$\frac{\tau_{\phi,0}}{\rho\Omega^2 b^2} = -\frac{1}{\pi}\frac{\Omega'}{\Omega} Re_{\phi}^{-1} \frac{\dot{m}_0}{\mu r}, \qquad (2.56)$$

where Re_{ϕ} is the rotational Reynolds number (see list of symbols). (The final ratio on the right-hand side is the local nondimensional mass flow-rate.)

2.3 INTEGRAL EQUATIONS FOR A SINGLE DISC

Where the boundary-layer equations (2.44) to (2.48) are valid, it is possible to reduce them to a set of ordinary differential equations by integrating with respect to z across the boundary layer. This layer is assumed to be of finite thickness δ which, in general, is a function of r. The boundary conditions used for the integration are

$$v_r(0) = 0,\ v_{\phi}(0) = v_{\phi,0},\ v_z(0) = 0,\ T(0) = T_0,\ \tau(0) = \tau_0, \qquad (2.57)$$

$$v_r \to 0,\ v_{\phi} \to v_{\phi,\infty},\ v_z \to v_{z,\infty},\ T \to T_{\infty},\ \tau \to 0 \text{ when } z = \delta. \qquad (2.58)$$

(For a rotating disc, $v_{\phi,0} = \Omega r$; for a stationary disc, $v_{\phi,0} = 0$.)

The price to be paid for the simplicity of manipulation is that assumptions must be made about the way in which the dependent variables depend on z within the boundary layer. The integrated equations give the dependence on r of average values (the average being across the layer) of the dependent variables.

2.3.1 Momentum-integral equations

Integration of equations (2.44) to (2.46) between $z = 0$ and $z = \delta$ gives

$$\frac{d}{dr}\left(r\int_0^{\delta} v_r\, dz\right) + r v_{z,\infty} = 0, \qquad (2.59)$$

$$\frac{d}{dr}\left(r\int_0^{\delta} v_r^2\, dz\right) - \int_0^{\delta}(v_{\phi}^2 - v_{\phi,\infty}^2)\, dz = -\frac{r}{\rho}\tau_{r,0}, \qquad (2.60)$$

$$\frac{d}{dr}\left(r^2\int_0^\delta v_r v_\varphi \, dz\right) + r^2 v_{\varphi,\infty} v_{z,\infty} = -\frac{r^2}{\rho}\tau_{\varphi,0}, \tag{2.61}$$

and equation (2.59) can be used to eliminate $v_{z,\infty}$ from equation (2.61). Further, the mass flow-rate within the boundary layer is

$$\dot{m}_0 = 2\pi r\rho\int_0^\delta v_r \, dz, \tag{2.62}$$

so that equation (2.59) gives

$$v_{z,\infty} = -\frac{1}{2\pi\rho r}\frac{d\dot{m}_0}{dr}. \tag{2.63}$$

Equations (2.60) and (2.61) can then be rearranged in the form

$$\frac{d}{dr}\left(r\int_0^\delta v_r^2 \, dz\right) - \int_0^\delta (v_\varphi^2 - v_{\varphi,\infty}^2) \, dz = -\frac{r}{\rho}\tau_{r,0}, \tag{2.64}$$

$$\frac{d}{dr}\left(r^2\int_0^\delta v_r(v_\varphi - v_{\varphi,\infty}) \, dz\right) + \frac{\dot{m}_0}{2\pi\rho}\frac{d}{dr}(r v_{\varphi,\infty}) = -\frac{r^2}{\rho}\tau_{\varphi,0}. \tag{2.65}$$

Many authors omit the second term on the left-hand side of equation (2.61) (see Section 6.3.1). This is justified only if either there is no axial flow out of or into the boundary layer or there is no swirl outside the boundary layer ($v_{\varphi,\infty} = 0$); in general neither of these conditions is valid in a rotor-stator system.

The method is often implemented by assuming similarity solutions of the form

$$v_r = \alpha(r)[v_{\varphi,0}(r) - v_{\varphi,\infty}(r)]f(\zeta), \quad 0 \le \zeta \le 1, \tag{2.66}$$

$$v_\varphi = v_{\varphi,0}(r) - [v_{\varphi,0}(r) - v_{\varphi,\infty}(r)]g(\zeta), \quad 0 \le \zeta \le 1, \tag{2.67}$$

where $\alpha(r)$ is a function to be determined (and is often assumed to be a constant) and ζ is a nondimensional axial variable given by

$$\zeta = \frac{z}{\delta}. \tag{2.68}$$

The functions $f(\zeta)$, $g(\zeta)$ are chosen to give good approximations to the true profiles of v_r and v_ϕ respectively for $0 \leq z \leq \delta$; the boundary conditions (2.57), (2.58) imply that

$$f(0) = f(1) = g(0) = 0, \; g(1) = 1. \tag{2.69}$$

The form of the functions $v_{\phi,0}(r)$, $v_{\phi,\infty}$ depends on the problem under discussion. In most cases $v_{\phi,0}(r)$ is known; for a single disc in an infinite fluid it is assumed that $v_{\phi,\infty}$ is also known. It is also usually assumed that

$$\tau_{r,0} = -\alpha\,\tau_{\phi,0} \tag{2.70}$$

The moment, M, on one side of the disc may be obtained by integrating equation (2.65) with respect to r from the inner radius, a, to the outer radius, b, of the disc. This gives

$$M = -2\pi \int_a^b r^2 \tau_{\phi,0}\, \mathrm{dr} = \int_a^b \dot{m}_0 \frac{d}{dr}(r v_{\phi,\infty})\, dr + K_v Z, \tag{2.71}$$

where K_v is the *velocity shape factor* defined by

$$K_v = \frac{\int_0^\delta v_r (v_\phi - v_{\phi,\infty}) dz}{(v_{\phi,0} - v_{\phi,\infty}) \int_0^\delta v_r dz} \tag{2.72}$$

and Z is an "angular-momentum deficit" defined by

$$Z = \Big[b\Big(v_{\phi,0}(b) - v_{\phi,\infty}(b)\Big)\dot{m}_0(b) - a\Big(v_{\phi,0}(a) - v_{\phi,\infty}(a)\Big)\dot{m}_0(a)\Big]. \tag{2.73}$$

For the important case in which there is solid-body rotation outside the boundary layer, it follows from equation (2.65) that there is a relationship (similar to equation (2.56) derived for Ekman layers) between the nondimensional shear stress on the disc, the rotational Reynolds number and the nondimensional local mass flow-rate. This is

$$\frac{\tau_{\phi,0}}{\rho\Omega^2 b^2} = -\frac{1}{\pi} K_w Re_\phi^{-1} \frac{\dot{m}_0}{\mu r}, \qquad (2.74)$$

where K_w is a *stress-flow-rate parameter* defined by

$$K_w = \left(1 + \frac{1}{2}\frac{r}{\dot{m}_0}\frac{d\dot{m}_0}{dr}\right)K_v \frac{(v_{\phi,0} - v_{\phi,\infty})}{\Omega r} + \frac{v_{\phi,\infty}}{\Omega r}, \qquad (2.75)$$

and K_v is given by equation (2.72). It is of interest to note that as $v_{\phi,\infty} \to v_{\phi,0}$ (so that the fluid is in near solid-body rotation), equation (2.75) shows that $K_w \to 1$; this is consistent with the result found in equation (2.56) when $\Omega' = \Omega$.

For similarity solutions of the form given in equations (2.66) and (2.67), the expression (2.72) for the shape factor reduces to

$$K_v = \frac{\int_0^1 f(\zeta)[1 - g(\zeta)]d\zeta}{\int_0^1 f(\zeta)d\zeta}, \qquad (2.76)$$

and this is independent of radius. In this case, it usual to assume that $\dot{m}_0/\mu r = \epsilon_m (x^2 Re_\phi)^\sigma$, where $x = r/b$ and σ is constant; ϵ_m and σ are referred to as *similarity parameters*. The first bracket in equation (2.75) then simplifies to $(\sigma + \frac{3}{2})$.

Various expressions may be used for the functions $f(\zeta)$ and $g(\zeta)$. For turbulent flow, the most common is the so-called $\frac{1}{7}$ - power-law profile in which

$$f(\zeta) = \zeta^{\frac{1}{7}}(1 - \zeta), \quad g(\zeta) = \zeta^{\frac{1}{7}}, \qquad (2.77)$$

and it is appropriate to consider this case here. The expressions for the components of stress at the wall are

$$\tau_{r,0} = \tag{2.78}$$
$$0.0225\rho\left(\frac{\mu}{\rho\delta}\right)^{\frac{1}{4}} \operatorname{sgn}(v_{\phi,0} - v_{\phi,\infty}) |v_{\phi,0} - v_{\phi,\infty}|^{\frac{7}{4}}\alpha(1 + \alpha^2)^{\frac{3}{8}},$$

$$\tau_{\phi,0} = \tag{2.79}$$
$$-0.0225\rho\left(\frac{\mu}{\rho\delta}\right)^{\frac{1}{4}} \operatorname{sgn}(v_{\phi,0} - v_{\phi,\infty}) |v_{\phi,0} - v_{\phi,\infty}|^{\frac{7}{4}}(1 + \alpha^2)^{\frac{3}{8}}.$$

Equations (2.62), (2.64) and (2.65) become

$$\dot{m}_0 = \frac{49\pi}{60}\rho r\alpha(v_{\phi,0} - v_{\phi,\infty})\delta, \tag{2.80}$$

$$\frac{343}{1656}\frac{d}{dr}\left[r\alpha^2(v_{\phi,0} - v_{\phi,\infty})^2\delta\right] - \left[\frac{1}{8}(v_{\phi,0}^2 - v_{\phi,\infty}^2) - \frac{7}{72}(v_{\phi,0} - v_{\phi,\infty})^2\right]\delta = -\frac{r}{\rho}\tau_{r,0}, \tag{2.81}$$

$$\frac{49}{720}\frac{d}{dr}\left[r^2\alpha(v_{\phi,0} - v_{\phi,\infty})^2\delta\right] + \frac{\dot{m}_0}{2\pi\rho}\frac{d}{dr}(rv_{\phi,\infty}) = -\frac{r^2}{\rho}\tau_{\phi,0}, \tag{2.82}$$

where $\tau_{r,0}$, $\tau_{\phi,0}$ are given by equations (2.78) and (2.79).

2.3.2 Ekman integral equations

A similar procedure can be carried out for the Ekman-layer equations discussed in section 2.2.2. The resulting integral equations are

$$-2\Omega'\int_0^\delta (v - v_\infty)\,dz = -\frac{1}{\rho}\tau_{r,0} \tag{2.83}$$

$$2\Omega'\int_0^\delta u\,dz = -\frac{1}{\rho}\tau_{\phi,0}, \tag{2.84}$$

where $v_\infty = v_{\phi,\infty} - \Omega' r$ is the value of v outside the Ekman layer. The value of Ω' used varies according to the problem under consideration, and it is frequently useful to replace Ω' by Ω, the angular speed of the disc. (The application of the Ekman integral equations to the flow in rotating cavities is discussed in detail by Owen, Pincombe and Rogers 1985.)

2.3.3 Energy-integral equation

In general the thicknesses, δ and δ_T, of the thermal and the velocity boundary layers respectively are not equal and the energy equation (2.49) must be integrated from $z = 0$ to $z = \delta'$ where δ' is the larger of δ and δ_T. After integration, the equation can be written in the form

$$\frac{d}{dr}\left(r\int_0^{\delta'} v_r(H - H_\infty)dz\right) + \frac{dH_\infty}{dr}\int_0^{\delta'} rv_r dz = \frac{r}{\rho}(q_0 - v_{\phi,0}\tau_{\phi,0}). \quad (2.85)$$

If $\delta_T > \delta$, then $v_r = 0$ when $\delta < z < \delta_T$ and δ' $(= \delta_T)$ can be replaced by δ; if $\delta_T < \delta$, then $\delta' = \delta$. In either case, the upper limit of the two integrals on the left-hand side of equation (2.85) can be replaced by δ, and the equation becomes

$$\frac{d}{dr}\left(r\int_0^{\delta} v_r(H - H_\infty)\, dz\right) + \frac{\dot{m}_0}{2\pi\rho}\frac{dH_\infty}{dr} = \frac{r}{\rho}(q_0 - v_{\phi,0}\tau_{\phi,0}). \quad (2.86)$$

When viscous dissipation is negligible, equation (2.86) reduces to

$$\frac{d}{dr}\left(r\int_0^{\delta} v_r(T - T_\infty)\, dz\right) + \frac{\dot{m}_0}{2\pi\rho}\frac{dT_\infty}{dr} = \frac{rq_0}{\rho c_p}. \quad (2.87)$$

The energy-integral equation is not, in general, dealt with in the same way as the momentum-integral equations because the enthalpy (or temperature) profile depends both on the Prandtl number and on the disc-temperature distribution. Discussion of the use of the equations is deferred to Chapter 5.

2.4 MOMENTUM-INTEGRAL EQUATIONS FOR A ROTOR-STATOR SYSTEM

2.4.1 Large clearance between rotor and stator

It often happens that the flow between the rotor and the stator consists of an inviscid core together with separate boundary layers on the two discs. It may be assumed that the equations derived in Section 2.4.1 are applicable to both boundary layers with $v_{\phi,0} = \Omega r$ for the rotor and $v_{\phi,0} = 0$ for the stator. There are essentially five unknowns: these are $\alpha_0(r)$, $\delta_0(r)$, $\alpha_s(r)$, $\delta_s(r)$ and $v_{\phi,c}(r)$, where the subscripts 0 and s are used here to denote values on the rotor and stator respectively. It is assumed that $v_{\phi,c}(r)$ is independent of z, as discussed in Section 2.2. (Care must be taken over the sign of $v_{\phi,c}(r)$, when using the equations.)

So far four momentum equations (two for each disc) have been derived to determine these five unknowns. The fifth relationship is obtained from continuity considerations; when there is no radial flow in the core it follows that, for steady flow,

$$\dot{m}_0 + \dot{m}_s = \dot{m}, \tag{2.88}$$

where $\dot{m}$ is independent of r (and takes the value zero when there is no superposed radial flow).

It is of importance to the designer to be able to calculate the frictional moment exerted by the rotor on the fluid. If it is assumed that the form of the velocity profiles f and g are the same for the two discs, equation (2.71) holds with the same value of K_v for both the rotor and the stator. (The equation is correct for each disc since the signs of $v_{\phi,c}$ and M are reversed and $v_{\phi,0}$ is zero on the stator.) It follows, after some manipulation, that

$$M_0 + M_s = \tag{2.89}$$

$$(1 - K_v)\dot{m}[bv_{\phi,c}(b) - av_{\phi,c}(a)] + K_v\,[\Omega b^2\dot{m}_0(b) - \Omega a^2\dot{m}_0(a)].$$

In the special case in which $f(\eta)$ and $g(\eta)$ are given by equation (2.77), the shape factor $K_{\upsilon} = \frac{1}{6}$.

This equation takes no account of the moment on the peripheral shroud. The first term on the right-hand side is the rate of loss of angular momentum in the cavity due to a superposed radial flow. The second term occurs since the fluid near the rotor must, of necessity, have more angular momentum near the periphery than near the axis.

2.4.2 Small clearance between rotor and stator

When there is no inviscid core between a rotor and a stator, equations (2.35) to (2.37) may be integrated with respect to z across the whole clearance. The boundary conditions are

$$v_r = 0,\ v_\phi = \Omega r,\ v_z = 0,\ \tau = \tau_0 \quad \text{when } z = 0, \tag{2.90}$$

$$v_r = 0,\ v_\phi = 0,\ v_z = 0,\ \tau = \tau_s \quad \text{when } z = s. \tag{2.91}$$

Equations (2.59) to (2.61) are replaced by

$$\frac{d}{dr}\left(r\int_0^s v_r\, dz\right) = 0, \tag{2.92}$$

$$\frac{d}{dr}\left(r\int_0^s v_r^2\, dz\right) - \int_0^s v_\phi^2\, dz = \frac{r}{\rho}(\tau_{r,s} - \tau_{r,0}), \tag{2.93}$$

$$\frac{d}{dr}\left(r^2\int_0^s v_r v_\phi\, dz\right) = \frac{r^2}{\rho}(\tau_{\phi,s} - \tau_{\phi,0}). \tag{2.94}$$

2.5 FRICTION AND HEAT TRANSFER: THE REYNOLDS ANALOGY

The Reynolds analogy between heat transfer and wall-stress for a fluid with a Prandtl number, Pr, of unity, flowing over a flat plate (see Schlichting 1979) is well known. A similar result is true for the tangential component of stress and the heat transfer on a rotating disc. (Owing to the presence of the pressure-gradient term in equation (2.36), there is no corresponding analogy associated with the radial component of stress.)

The application of the Reynolds analogy to rotating flows was given by Dorfman (1963) for the case in which viscous dissipation is negligible; his work was extended by Owen (1971) to include dissipative effects when the core rotation is zero, and Chew (1985) considered the case of nonzero core rotation. The discussion of the boundary conditions given here is an extension of the work of Owen and Chew.

2.5.1 Similarity between enthalpy and swirl

If the assumptions given in equations (2.51) are valid, and it is assumed that

$$\frac{k_{eff}}{c_p} = \mu_{r,eff} = \mu_{\phi,eff} = \mu_{eff} \text{ (say)}, \tag{2.95}$$

(this corresponds to $Pr = Pr_{turb} = 1$, where Pr_{turb} is defined in Section 2.2.1), equation (2.49) can be written in the form

$$\frac{\partial}{\partial r}[rv_r(H - H_{ref})] + \frac{\partial}{\partial z}[rv_z(H - H_{ref})] = \frac{r}{\rho}\frac{\partial}{\partial z}\left[\mu_{eff}\frac{\partial(H - H_{ref})}{\partial z}\right], \tag{2.96}$$

where H_{ref} is a constant reference enthalpy appropriate to the problem under discussion.

Also, defining the swirl, Γ, as

$$\Gamma = rv_\phi, \tag{2.97}$$

equation (2.37) can be written as

$$\frac{\partial}{\partial r}(rv_r\Gamma) + \frac{\partial}{\partial z}(rv_z\Gamma) = \frac{r}{\rho}\frac{\partial}{\partial z}\left[\mu_{eff}\frac{\partial \Gamma}{\partial z}\right], \tag{2.98}$$

When $z = 0$, $\Gamma = \Gamma_0 = \Omega r^2$, and for the analogy to be valid it is necessary to take

$$H_0(r) - H_{ref} = C\Omega^2 r^2, \tag{2.99}$$

where C is a constant (the factor Ω^2 is introduced to make C nondimensional) and $H_0(r)$ is the value of H when $z=0$. In terms of the temperature $T_0(r)$ on the disc, this is equivalent to a "quadratic" distribution of the form $T_0(r) = T_{ref} + K\Omega^2 r^2/c_p$, where T_{ref} is a reference temperature (defined by $H_{ref} = c_p T_{ref}$) and K is a nondimensional constant equal to $C - \frac{1}{2}$. It follows that

$$T_0 - T_{ref} = (C - \tfrac{1}{2})\frac{\Omega^2 r^2}{c_p} \tag{2.100}$$

and that

$$T_{ref} = T_{0,a} - (C - \tfrac{1}{2})\frac{\Omega^2 a^2}{c_p}, \tag{2.101}$$

where $T_{0,a}$ is the disc temperature where $r = a$. Such quadratic distributions are often reasonable approximations to those that occur in practical cases.

The second boundary condition is defined at the edge of the boundary layer as $z \to \infty$. Since both $v_{\phi,\infty}$ and H_∞ are independent of z, it follows that

$$\left[\frac{\partial \Gamma}{\partial z}\right]_\infty = 0, \quad \left[\frac{\partial H}{\partial z}\right]_\infty = 0. \tag{2.102}$$

The condition for similarity at this boundary is automatically satisfied. It should be pointed out that previous authors have usually required, instead of the condition (2.102), that $H_\infty(r) - H_{ref} = C\Omega\Gamma_\infty$.

Strictly, the initial conditions on Γ and $H - H_{ref}$ at or near $r = a$ (the start of the boundary layer) must also be similar. However, in most boundary-layer solutions the effect of the initial conditions decays rapidly with increasing r. Even if the conditions are not similar near $r = a$, the analogy will become increasingly accurate as r increases.

Under these conditions, the distribution of Γ and $H - H_{ref}$ will be similar throughout the boundary layer and

$$H - H_{ref} = C\Omega\Gamma \tag{2.103}$$

everywhere. (In particular, as $z \to \infty$, equation (2.103) gives the alternative boundary condition to equation (2.102).) It is convenient to express this similarity in several ways, two of which are given here:

$$\frac{H - H_{ref}}{H_0 - H_{ref}} = \frac{\Gamma}{\Gamma_0}, \tag{2.104}$$

$$\frac{H - H_\infty}{H_0 - H_\infty} = \frac{\Gamma - \Gamma_\infty}{\Gamma_0 - \Gamma_\infty}. \tag{2.105}$$

In particular equation (2.104) is valid as $z \to \infty$, and so

$$\frac{H_\infty - H_{ref}}{H_0 - H_{ref}} = \frac{\Gamma_\infty}{\Gamma_0}. \tag{2.106}$$

It is worth emphasizing that no assumption has been made about the relationship between the enthalpy and the circulation outside the boundary layer: equation (2.106) is a *consequence* of the boundary condition (2.102).

From equation (2.105) it follows that H_∞ (and, hence, T_∞) and Γ_∞ cannot both be specified: if one is known, then the other will adjust itself so that equation (2.105) holds. (An interdependence of H_∞ and Γ_∞ would also be expected to occur even if $Pr \neq 1$ or if the temperature distributions are not quadratic.) In particular, it is clear from equation (2.105) that H_∞ cannot be constant unless either $\Gamma_\infty = 0$ (in which case the fluid outside the boundary layer is quiescent) or Γ_∞ is constant (in which case the flow outside the boundary layer is a free vortex).

When dissipative effects are negligible, equations (2.104) to (2.106) are valid with H replaced by T everywhere.

The analogy, which is applied to a rotating disc, needs little modification for use with a stationary disc. Since there is no swirl on the stator itself, the analogy is valid only for an isothermal disc: equation (2.99) then gives $H_0 = H_{ref}$ (here the subscript 0 refers, exceptionally, to the stator) and the analogy still gives $H - H_{ref} \propto \Gamma$. Equations (2.104) and (2.106) are meaningless for the isothermal stationary disc, since all the denominators are zero; equation (2.105) is still valid.

2.5.2 Relationship between heat flux and stress

From the definition of H given by equation (2.48), and using equations (2.51) and (2.95), the heat flux q can be calculated from

$$q = -\mu_{eff}\frac{\partial}{\partial z}(c_p T) = -\mu_{eff}\frac{\partial H}{\partial z} + v_r\tau_r + v_\phi\tau_\phi. \qquad (2.107)$$

For the case where the Reynolds analogy is valid and conditions in the core of fluid outside the boundary layer are known, it is convenient to use equation (2.105) together with equation (2.107) to give

$$q = -\frac{H - H_\infty}{\Gamma_0 - \Gamma_\infty} r\tau_\phi + v_r\tau_r + v_\phi\tau_\phi. \qquad (2.108)$$

On the disc, $v_r = 0$, $v_\phi = \Omega r$, $\tau_\phi = \tau_{\phi,0}$ and $q = q_0$; hence

$$q_0 = -\frac{\tau_{\phi,0}}{\rho\Omega^2 b^2} Re_\phi \frac{\mu c_p (T_0 - T_\infty)}{r(1-\beta)} [1 - \tfrac{1}{2} Ec(1-\beta)^2], \qquad (2.109)$$

where Ec is the Eckert number based on the temperature difference $T_0 - T_\infty$ (see list of symbols) and $\beta = v_{\phi,\infty}/\Omega r$ is the *core-swirl ratio*.

Alternatively, equation (2.104) can be used with equation (2.108) to evaluate the heat transfer. This gives

$$q_0 = -\frac{\tau_{\phi,0}}{\rho\Omega^2 b^2} Re_\phi \frac{\mu c_p (T_0 - T_{ref})}{r} [1 - \tfrac{1}{2} Ec_{ref}], \qquad (2.110)$$

where Ec_{ref} is the Eckert number based on the temperature difference $T_0 - T_{ref}$. This form is simpler to use than equation (2.109) as long as T_{ref} is known. Using the expression for $T_0 - T_{ref}$ given in equation (2.100), it can be seen that

$$q_0 = (C - 1)\Omega r \tau_{\phi,0}. \qquad (2.111)$$

2.5.3 Adiabatic-disc temperature

If the disc is adiabatic, it is convenient to use a different set of boundary conditions. The condition on the disc is obtained by differentiating equation (2.48) with respect to z and putting $z = 0$; since $\partial p/\partial z$ is zero throughout the boundary layer, and $\partial T/\partial z$ is zero on the surface of an adiabatic disc, it follows that

$$\left[\frac{\partial H}{\partial z}\right]_{z=0} = \Omega \left[\frac{\partial \Gamma}{\partial z}\right]_{z=0}. \qquad (2.112)$$

For the analogy to hold, the condition outside the boundary layer is given by equation (2.102). As before the initial conditions at $r = a$ are assumed to be satisfied. With these conditions it follows that the distributions of Γ and $H - H_{ref}$ are similar; hence

$$H_{ad} - H_{ref} = \Omega\Gamma \qquad (2.113)$$

everywhere in the boundary layer, where the subscript *ad* indicates the value for an adiabatic disc. This equation is identical with equation (2.103) for $C = C_{ad} = 1$. The adiabatic-disc temperature, $T_{0,ad}$, may be deduced from equation (2.113) as

$$T_{0,ad} = T_{ref} + \frac{\Omega^2 r^2}{2c_p}. \qquad (2.114)$$

Equations (2.113) and (2.114) could have been deduced from equation (2.111) and (2.110) respectively, but this would have involved the assumption that $T_{0,ad}$ is quadratic. The derivation given here implies that the adiabatic-disc temperature is

(i) quadratic, and

(ii) independent of conditions in the core.

An alternative expression for $T_{0,ad}$ is obtained by putting q_0 equal to zero in equation (2.109), which still holds; then

$$T_{0,ad} = T_{\infty,ad} + \frac{\Omega^2 r^2}{2c_p}(1 - \beta)^2 . \qquad (2.115)$$

At first sight, this form appears to contradict the result that $T_{0,ad}$ is independent of conditions in the core. It must be remembered, however, that $T_{\infty,ad}$ depends on β (see equation (2.106)).

When dissipation is negligible, equations (2.114) and (2.115) reduce to $T_{0,ad} = T_{\infty,ad} = T_{ref}$.

2.5.4 Nusselt numbers and moment coefficients

When the disc is not adiabatic, the nondimensional heat flux can be expressed in terms of a local Nusselt number, Nu and an average Nusselt number, Nu_{av}, (see list of symbols). For use with the Reynolds analogy, the heat flux q_w (used in the definition of Nu) is clearly equal to q_0, and the temperature difference ΔT can be taken as $T_0 - T_{0,ad}$, where $T_{0,ad}$ is given by equation (2.114). It is convenient to use an asterisk to denote the Nusselt numbers using this special value of ΔT. Then

$$Nu^* = \frac{rq_0}{k(T_0 - T_{0,ad})}, \quad Nu^*_{av} = \frac{bq_{0,av}}{k(T_0 - T_{0,ad})_{av}}, \qquad (2.116)$$

where $q_{0,av}$ and $(T_0 - T_{0,ad})_{av}$ are radially-weighted averages.

Equation (2.110) can be used to give

$$Nu^* = -Re_\phi \frac{\tau_{\phi,0}}{\rho\Omega^2 b^2}, \qquad (2.117)$$

and, using equation (2.74), this gives

$$Nu^* = \frac{1}{\pi} K_w \frac{\dot{m}_0}{\mu r}, \qquad (2.118)$$

where the stress-flow-rate parameter K_w, given by equation (2.75), is constant for many important flows. The average Nusselt number can be expressed as

$$Nu^*_{av} = \frac{Re_\phi C_M}{\pi}, \qquad (2.119)$$

where C_M is the moment coefficient (see list of symbols).

Thus, when the Prandtl number is unity and the disc-temperature distribution is quadratic, the local Nusselt number can be determinined directly from either the local shear stress or the local mass flow-rate, and the average Nusselt number from the moment coefficient. For this special case there is, therefore, no need to solve the energy equation.

2.5.5 Modification to the Reynolds analogy

When the Prandtl number is not unity or the disc-temperature distribution is not quadratic, it is often convenient to introduce factors into the expressions obtained above for $T_{0,ad}$ and for q_0. Different authors have suggested alternative ways of doing this. When $Pr \neq 1$, a *recovery factor* is usually introduced into one of the expressions for $T_{0,ad}$, so that equation (2.115) can be modified to

$$T_{0,ad} = T_{\infty,ad} + R \frac{\Omega^2 r^2}{2c_p} (1 - \beta)^2, \qquad (2.120)$$

and equation (2.114) to

$$T_{0,ad} = T_{ref} + R' \frac{\Omega^2 r^2}{2c_p}. \qquad (2.121)$$

The recovery factors R and R' both depend on Pr and give, in general, different results for the adiabatic-disc temperature. (Both probably also depend on conditions far from the disc and on the disc-temperature distribution: these effects are usually neglected in practice.)

When the disc is not adiabatic, the heat transfer is sometimes determined using correction factors, χ and χ', such that

$$q_0 = -\chi \frac{\tau_{\phi,0}}{\rho\Omega^2 b^2} Re_\phi \frac{\mu c_p (T_0 - T_\infty)}{r(1-\beta)} [1 - \tfrac{1}{2}Ec\, R\,(1-\beta)^2], \quad (2.122)$$

where, as for equation (2.109), Ec is the Eckert number based on the temperature difference $T_0 - T_\infty$ and $\beta = v_{\phi,\infty}/\Omega r$. Alternatively, the heat transfer can be expressed as

$$q_0 = -\chi' \frac{\tau_{\phi,0}}{\rho\Omega^2 b^2} Re_\phi \frac{\mu c_p (T_0 - T_{ref})}{r} [1 - \tfrac{1}{2}Ec_{ref} R'], \quad (2.123)$$

which is related to equation (2.110).

It was shown in Section 2.5.1 that, in general, T_∞ depends on Γ_∞ (and, hence, on β). For the free disc (where $\beta = 0$, $\Gamma_\infty = 0$), T_∞ is constant when the Reynolds analogy is valid and this value can be used for T_{ref}; the same result is assumed to hold when $Pr \neq 1$ or the disc-temperature distribution is not quadratic. It follows from equations (2.122) and (2.123) that, for the free disc, $\chi' = \chi$ and $R' = R$.

When $\beta \neq 0$, any theory used to predict q_0 must also predict T_∞. It is interesting to note that equations (2.122) and (2.123) imply that

$$\frac{\chi'}{\chi} = \frac{1}{1-\beta} \frac{T_0 - T_\infty}{T_0 - T_{ref}} \frac{1 - \frac{1}{2} Ec R\,(1-\beta)^2}{1 - \frac{1}{2} Ec_{ref} R'}. \quad (2.124)$$

It is convenient to consider the Nusselt number defined in terms of a temperature difference $T_0 - T_1$, where T_1 is a "suitable" reference temperature. For example, it is common to take $T_1 = T_\infty$ for the free disc; $T_1 = T_c$ (the temperature outside the boundary layer on the disc) for an enclosed disc; or $T_1 = T_I$ for a rotor-stator system with a superposed flow of fluid whose incoming temperature is T_I. Using equation (2.122), the Nusselt number can then be written as

$$Nu = -Pr\chi Re_\phi \frac{1}{1-\beta}\frac{\tau_{\phi,0}}{\rho\Omega^2 b^2}\frac{T_0 - T_\infty}{T_0 - T_1}[1 - \tfrac{1}{2}EcR(1-\beta)^2] \qquad (2.125)$$

or as

$$Nu = -Pr\chi' Re_\phi \frac{\tau_{\phi,0}}{\rho\Omega^2 b^2}\frac{T_0 - T_{ref}}{T_0 - T_1}[1 - \tfrac{1}{2}Ec_{ref}R']. \qquad (2.126)$$

When similarity solutions are appropriate, it is often convenient to use equation (2.74) to write these equations in the form

$$Nu = Pr\chi \frac{K_w}{\pi(1-\beta)}\frac{\dot{m}_0}{\mu r}\frac{T_0 - T_\infty}{T_0 - T_1}[1 - \tfrac{1}{2}EcR(1-\beta)^2] \qquad (2.127)$$

and

$$Nu = Pr\chi' \frac{K_w}{\pi}\frac{\dot{m}_0}{\mu r}\frac{T_0 - T_{ref}}{T_0 - T_1}[1 - \tfrac{1}{2}Ec_{ref}R'], \qquad (2.128)$$

where the stress-flow-rate parameter, K_w, is given by equation (2.74).

When dissipative effects are negligible, equations (2.122) to (2.128) are valid if $Ec = Ec_{ref} = 0$. When dissipation is taken into account, it is often more useful to use the Nusselt number Nu^*, defined using the temperature difference $\Delta T = T_0 - T_{0,ad}$. This is equivalent to taking $T_1 = T_{0,ad}$ in equations (2.127) and (2.128). It is easy to verify that

$$Nu^* = Pr\,\chi \frac{K_w}{\pi(1-\beta)}\frac{\dot{m}_0}{\mu r}\left(1 - \frac{T_\infty - T_{\infty,ad}}{T_0 - T_{0,ad}}\right) \qquad (2.129)$$

if the form (2.127) is used, or

$$Nu^* = Pr\,\chi' \frac{K_w}{\pi}\frac{\dot{m}_0}{\mu r} \qquad (2.130)$$

if the form (2.128) is used. The ratio of Nu to Nu^* is given by

$$\frac{Nu}{Nu^*} = \frac{T_0 - T_{0,ad}}{T_0 - T_{ref}} = 1 - \tfrac{1}{2}Ec_b R' x^2 \tag{2.131}$$

where Ec_b is the value of Ec_{ref} when $x = 1$ $(r = b)$. It is clear from this equation that the difference between Nu and Nu^* increases as Ec_b increases. When the Reynolds analogy holds ($Pr = 1$, $T_0 - T_{ref} \propto x^2$), Nu^* is independent of Ec_b (see equation (2.117)) and $R' = 1$; in this case, therefore, equation (2.131) shows that Nu depends on Ec_b: indeed when $Ec_b x^2 > 2$, Nu becomes negative! A similar result is expected to hold approximately in more general cases; this will be demonstrated in the theory of Section 5.3.3 for the free disc. It is strongly recommended, therefore, that Nu^*, rather than Nu, should be used as the appropriate Nusselt number in practical applications. (When Ec_b is small, there is little difference between Nu and Nu^*.)

It can be shown that

$$Nu^*_{av} = Pr\, \chi_{av} \frac{Re_\phi C_M}{\pi}, \tag{2.132}$$

where χ_{av} is a suitable average. For the free disc with a quadratic temperature distribution, $\chi_{av} = \chi' = \chi$, and equation (2.132) becomes

$$Nu^*_{av} = Pr\, \chi \frac{Re_\phi C_M}{\pi}. \tag{2.133}$$

When $Pr = 1$, this reduces to equation (2.119).

The average Nusselt number, Nu_{av}, can be written as a quadrature involving the distribution of Nu. When both are defined using the temperature difference $T_0 - T_1$, it is easy to demonstrate that

$$Nu_{av} = \frac{b\int_0^b (T_0 - T_1) Nu\, dr}{\int_0^b r(T_0 - T_1) dr}. \tag{2.134}$$

A particular form of this equation is, of course,

$$Nu^*_{av} = \frac{b\int_0^b (T_0 - T_{0,ad})Nu^*\, dr}{\int_0^b r(T_0 - T_{0,ad})dr} . \qquad (2.135)$$

A differential form of equation (2.134) can be found if a "running average" is used. An alternative average Nusselt number, $\overline{Nu}$ is defined taking $q_w = \overline{q_0}$ and $\Delta T = \overline{(T_0 - T_1)}$ and replacing b in the definition by r. The overbars for the heat flux and the temperature difference indicate radially-weighted averages over the disc up to radius r. Thus

$$\overline{Nu} = \frac{r\int_0^r rq_0 dr}{k\int_0^r r(T_0 - T_1)dr} . \qquad (2.136)$$

It is then possible to show that

$$Nu = \overline{Nu} + \left[\frac{1}{T_0 - T_1}\int_0^r r(T_0 - T_1)dr\right]\frac{d}{dr}\left(\frac{\overline{Nu}}{r}\right). \qquad (2.137)$$

CHAPTER 3
Laminar Flow over a Single Disc

Before investigating the complicated flow that occurs in rotor-stator systems, it is useful to discuss the flow due to a single disc rotating in free space. Four cases are of interest: in the first, which is often referred to as *the free disc*, there is no flow far from the disc; in the second, the fluid far from the disc is rotating at a speed less than that of the disc; in the third, the fluid far from the disc is rotating and the disc itself is stationary; and in the fourth, an axial flow is superimposed on the rotating disc.

The disc, of radius b, is in the plane $z = 0$ and rotates with angular velocity Ω about the z-axis (see Figure 3.1). A boundary layer on each side of the disc develops, through which the tangential component of velocity, v_φ, is sheared from the value Ωr at the surface to the value zero in the "free stream" outside the boundary layer. In practice, only the flow on one side of the disc ($z > 0$) need be discussed, and only the region for which $r < b$.

The "centrifugal forces" created by the rotating disc cause a radial outflow of fluid within the boundary layer. Since the radial component of velocity is zero both on the disc and in the free stream, external fluid is entrained axially into the boundary layer; the radial outflow is often referred to as the "free-disc pumping effect".

Inside the boundary layer, the flow may be laminar or turbulent: the determining parameter is the local rotational Reynolds number $x^2 Re_\varphi = \Omega r^2/\nu$. For all values of Ω, the flow near the axis of rotation is usually laminar; for large values of Ω, the flow may become turbulent, even for comparatively small values of r. In the latter case, there will be a transition region between the laminar and turbulent flow. The three types of flow are discussed separately below.

3.1 LAMINAR FLOW OVER THE FREE DISC

3.1.1 Theory

For the region $z > 0$, $r < b$ (shown on the right-hand side of the disc in Figure 3.1), the boundary conditions for the velocity components are

$$v_r = 0,\ v_\varphi = \Omega r,\ v_z = 0 \quad \text{at } z = 0, \tag{3.01}$$

$$v_r \to 0,\ v_\varphi \to 0 \quad \text{as } z \to \infty. \tag{3.02}$$

The value of v_z as $z \to \infty$ is not specified; it adjusts to a nonzero negative value, which provides exactly the right amount of fluid necessary to maintain the pumping effect described above.

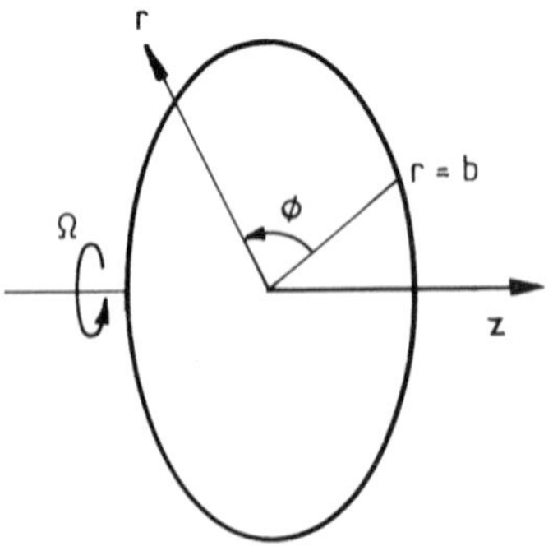

Figure 3.1
Coordinate system for a rotating disc

The problem was treated theoretically by von Kármán (1921), who defined the new independent variable

$$\zeta = z\left(\frac{\Omega}{\nu}\right)^{\frac{1}{2}}, \tag{3.03}$$

and assumed that the velocity components and the pressure are of the form

$$\begin{aligned} v_r = \Omega r F(\zeta),\ v_\phi = \Omega r G(\zeta),\ v_z &= (\Omega\nu)^{\frac{1}{2}} H(\zeta), \\ p &= -\rho\Omega\nu P(\zeta). \end{aligned} \tag{3.04}$$

Using these expressions in equations (2.03) to (2.06), it is found that

$$H' = -2F, \tag{3.05}$$

$$F'' = F^2 - G^2 + F'H, \tag{3.06}$$

$$G'' = 2FG + G'H, \tag{3.07}$$

$$P' = HH' - H'', \tag{3.08}$$

where primes indicate differentiation with respect to ζ. The boundary conditions for the first three of these equations (which are often referred to as von Kármán's equations) are derived from equations (3.01) and (3.02), and become

$$F(0) = 0,\ G(0) = 1,\ H(0) = 0, \tag{3.09}$$

$$F(\zeta) \to 0,\ G(\zeta) \to 0 \quad \text{as } \zeta \to \infty. \tag{3.10}$$

It may be noted that the last equation, (3.08), is the only one involving $P(\zeta)$; it may be integrated directly to give

$$P(\zeta) = P(0) + \frac{1}{2}H^2(\zeta) + 2F(\zeta). \tag{3.11}$$

The problem reduces, therefore, to finding the solution of equations (3.05) to (3.07) subject to the boundary conditions (3.09) and (3.10).

Before discussing techniques which may be used to solve the equations, it is useful to investigate the behaviour of their solution as $\zeta \to \infty$. Since $F(\zeta) \to 0$ as $\zeta \to \infty$, equation

(3.05) implies that $H \to H_\infty = -c$, a constant, as $\zeta \to \infty$. (The value of the constant c can be determined only by solution of the equations over the whole range of ζ; the minus sign is introduced, by hindsight, so that c is positive.) Assuming that $F^2(\zeta)$, $G^2(\zeta)$, $F(\zeta)G(\zeta)$, $F'(\zeta)[H(\zeta)+c]$ and $G'(\zeta)[H(\zeta)+c]$ are small quantities of the second order as $\zeta \to \infty$, equations (3.06) and (3.07) become

$$F'' + cF' \simeq 0, \qquad G'' + cG' \simeq 0. \tag{3.12}$$

It follows that, when $\zeta > \zeta_0$, where ζ_0 is large,

$$F(\zeta) \sim F(\zeta_0)e^{-c(\zeta-\zeta_0)}, \; G(\zeta) \sim G(\zeta_0)e^{-c(\zeta-\zeta_0)}, \; H \sim -c. \tag{3.13}$$

The full equations were first solved by Cochran (1934) who matched a series solution about $\zeta = 0$ with an asymptotic solution (for which equation (3.13) is the leading term) for large ζ. An alternative method, similar to that used by Rogers and Lance (1960), is to treat the problem as an initial value problem with a systematic series of "guessed" values for $F'(0)$ and $G'(0)$ chosen to ensure that equation (3.10) is satisfied.

Difficulties arise in the integration of the equations as they stand because of the infinite range of ζ. To determine $F'(0)$ and $G'(0)$ accurately, it is more satisfactory to modify the equations by using a transformation due to Benton (1966). A new independent variable, ξ, is defined as

$$\xi = e^{-c\zeta}, \tag{3.14}$$

and new dependent variables $f(\xi)$, $g(\xi)$ and $h(\xi)$ such that

$$F = c^2 f(\xi), \; G(\zeta) = c^2 g(\xi), \; H(\zeta) = -c[1 - h(\xi)]. \tag{3.15}$$

The range of ξ is from 0 (where $\zeta \to \infty$) to 1 (where $\zeta = 0$), and equations (3.05) to (3.07) may then be written as

$$\xi^2 \ddot{f} = f^2 - g^2 - \xi \dot{f} h \tag{3.16}$$

$$\xi^2 \ddot{g} = 2fg - \xi \dot{g} h \tag{3.17}$$

$$\xi \dot{h} = 2f, \tag{3.18}$$

where dots denote differentiation with respect to ξ. The boundary conditions (3.09) and (3.10), together with the extra condition that $h(0) = 0$, become

$$f(0) = g(0) = h(0) = 0, \tag{3.19}$$

$$f(1) = 0,\ g(1) = 1/c^2,\ h(1) = 1. \tag{3.20}$$

Five of these boundary conditions do not contain the unknown constant c, nor do equations (3.16) to (3.18); the system may therefore be solved uniquely. The constant c is then given by the sixth boundary condition in equation (3.20). Benton used power series solutions about $\xi = 0$ and $\xi = 1$ which he matched in the middle of the range; but the method presented here treats the problem as an initial value problem with guessed values of $\dot{f}(0)$ and $\dot{g}(0)$; corrected values are determined by systematic trial and error so that the conditions on $f(1)$ and $g(1)$ given by equation (3.20) are satisfied to the required accuracy. Since the general expressions for $\ddot{f}(\xi)$, $\ddot{g}(\xi)$ and $\dot{h}(\xi)$ are all indeterminate when $\xi = 0$, it is necessary to use a power-series expansion in ξ to determine $\ddot{f}(0)$, $\ddot{g}(0)$ and $\dot{h}(0)$; it is easy to show that $\ddot{f}(0) = -[\dot{f}^2(0) + \dot{g}^2(0)]$, $\ddot{g}(0) = 0$ and $\dot{h}(0) = 2\dot{f}(0)$. Equations (3.16) to (3.18) are then rewritten as five first-order equations in a way similar to that described in Appendix A for the solution of equations (3.05) and (3.07). Using a VAX 8530 computer with a typical precision of 16 significant figures, it was found that

$$\dot{f}(0) = 1.182245, \qquad \dot{g}(0) = 1.536777. \tag{3.21}$$

Once the equations have been solved, values of $H_\infty = -c$, $F'(0)$ and $G'(0)$ are evaluated using the relations

$$c = [1/g(1)]^{\frac{1}{2}}, \quad F'(0) = c^3\dot{f}(1), \quad G'(0) = c^3\dot{g}(1). \quad (3.22)$$

Although a considerably greater accuracy has been achieved it is, for most purposes, sufficient to take

$$H_\infty = -0.884474, \quad F'(0) = 0.510233, \quad G'(0) = -0.615922. \quad (3.23)$$

The solution (as a function of ζ) is tabulated in Table 3.1 and is shown graphically in Figure 3.2.

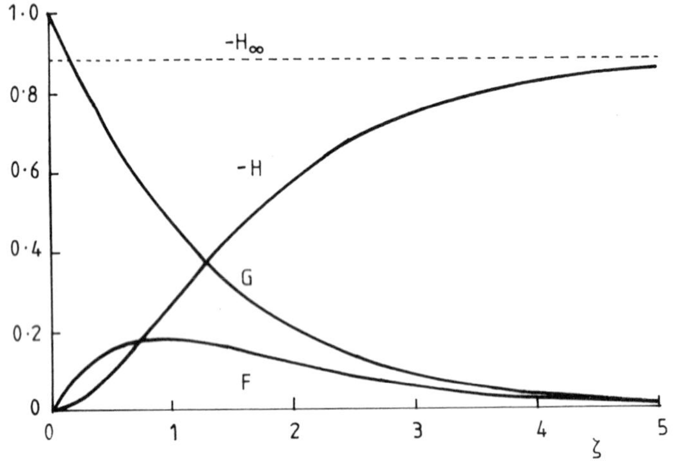

Figure 3.2 Laminar velocity profiles for the free disc

The solution given here agrees with that of Benton to four places of decimals for $\zeta \leq 4.4$ and for $\zeta \to \infty$ (with the exception of $F'(0.2)$, which is obviously a misprint in Benton's paper). Cochran's values are given to three places of decimals and, except for H_∞, his results are in perfect agreement with the more recent computations. The value of H_∞ obtained by Cochran was -0.886 which is in error by less than 0.2%.

If the boundary-layer thickness, δ, is defined to be the

Table 3.1 Numerical solutions of von Kármán's equations for the free disc

ζ	F	G	H	P-P(0)	F′	G′
0.0	0.0000	1.0000	0.0000	0.0000	0.5102	-0.6159
0.1	0.0462	0.9386	-0.0048	0.0925	0.4163	-0.6112
0.2	0.0836	0.8780	-0.0179	0.1674	0.3338	-0.5987
0.3	0.1133	0.8190	-0.0377	0.2274	0.2620	-0.5803
0.4	0.1364	0.7621	-0.0628	0.2747	0.1999	-0.5577
0.5	0.1536	0.7076	-0.0919	0.3115	0.1467	-0.5321
0.6	0.1660	0.6557	-0.1239	0.3396	0.1015	-0.5047
0.7	0.1742	0.6067	-0.1580	0.3608	0.0635	-0.4764
0.8	0.1789	0.5605	-0.1934	0.3764	0.0317	-0.4476
0.9	0.1807	0.5171	-0.2294	0.3877	0.0056	-0.4191
1.0	0.1802	0.4766	-0.2655	0.3955	-0.0157	-0.3911
1.1	0.1777	0.4389	-0.3013	0.4008	-0.0327	-0.3641
1.2	0.1737	0.4038	-0.3365	0.4041	-0.0461	-0.3381
1.3	0.1686	0.3712	-0.3707	0.4059	-0.0564	-0.3133
1.4	0.1625	0.3411	-0.4038	0.4066	-0.0640	-0.2898
1.5	0.1559	0.3132	-0.4357	0.4066	-0.0693	-0.2677
1.6	0.1487	0.2875	-0.4661	0.4061	-0.0728	-0.2470
1.7	0.1414	0.2638	-0.4952	0.4053	-0.0747	-0.2276
1.8	0.1338	0.2419	-0.5227	0.4043	-0.0754	-0.2095
1.9	0.1263	0.2218	-0.5487	0.4031	-0.0751	-0.1927
2.0	0.1189	0.2033	-0.5732	0.4020	-0.0739	-0.1771
2.1	0.1115	0.1864	-0.5962	0.4008	-0.0721	-0.1627
2.2	0.1045	0.1708	-0.6178	0.3998	-0.0698	-0.1494
2.3	0.0976	0.1565	-0.6380	0.3987	-0.0671	-0.1371
2.4	0.0910	0.1433	-0.6569	0.3978	-0.0643	-0.1258
2.5	0.0848	0.1313	-0.6745	0.3970	-0.0612	-0.1153
2.6	0.0788	0.1202	-0.6908	0.3962	-0.0580	-0.1057
2.7	0.0732	0.1101	-0.7060	0.3955	-0.0548	-0.0969
2.8	0.0678	0.1008	-0.7201	0.3949	-0.0517	-0.0888
2.9	0.0628	0.0923	-0.7332	0.3944	-0.0485	-0.0814
3.0	0.0581	0.0845	-0.7452	0.3939	-0.0455	-0.0745
3.2	0.0496	0.0708	-0.7668	0.3932	-0.0397	-0.0625
3.4	0.0422	0.0594	-0.7851	0.3926	-0.0343	-0.0524
3.6	0.0358	0.0498	-0.8007	0.3922	-0.0296	-0.0440
3.8	0.0304	0.0417	-0.8139	0.3919	-0.0253	-0.0369
4.0	0.0257	0.0349	-0.8251	0.3917	-0.0216	-0.0309
4.5	0.0168	0.0225	-0.8460	0.3914	-0.0144	-0.0199
5.0	0.0109	0.0144	-0.8596	0.3912	-0.0094	-0.0128
5.5	0.0070	0.0093	-0.8684	0.3912	-0.0062	-0.0082
6.0	0.0045	0.0060	-0.8741	0.3912	-0.0040	-0.0053
7.0	0.0019	0.0025	-0.8802	0.3912	-0.0017	-0.0022
8.0	0.0008	0.0010	-0.8827	0.3911	-0.0007	-0.0009
9.0	0.0003	0.0004	-0.8837	0.3911	-0.0003	-0.0004
10.0	0.0001	0.0002	-0.8842	0.3911	-0.0001	-0.0002
∞	0.0000	0.0000	-0.8845	0.3911	0.0000	0.0000

value of z for which $v_\varphi = 0.01\Omega r$ (that is, $G = 0.01$), it is easily seen from Table 3.1 that

$$\delta \simeq 5.5\left(\frac{\nu}{\Omega}\right)^{\frac{1}{2}}. \qquad (3.24)$$

Clearly δ is independent of r, a result that is implicit in the form of the solution assumed by von Kármán. That this is to be expected may be seen by discussing the orders of magnitude of the components τ_r, τ_φ of the shear stress. Within the boundary layer, the radial component of stress is balanced by the centrifugal force per unit area (which behaves like $\rho\Omega^2 r\delta$), and the tangential component is proportional to the axial shear of the tangential component of velocity (and so behaves like $\rho\Omega r/\delta$). Hence, the ratio of the two components of stress is

$$\frac{\tau_r}{\tau_\varphi} \propto \frac{\Omega\delta^2}{\nu}. \qquad (3.25)$$

If it is assumed that this ratio is independent of r (which is the case in practice), it follows that $\delta = K(\nu/\Omega)^{\frac{1}{2}}$ for some constant K. The actual value of K depends, of course, on the precise way in which δ is defined: for the definition given above, equation (3.19) gives $K \simeq 5.5$. An approximate solution given by Targ (1951), using momentum-integral techniques, gives $K = 3.50$, which corresponds to $v_\varphi \simeq 0.05\Omega r$.

The mass flow-rate $\dot{m}_0$, entrained by the boundary layer on one side of the disc, is equal to the axial flow-rate into the boundary layer from outside. It follows that

$$\dot{m}_0 = -2\pi\rho\int_0^r r\left[v_z\right]_{z\to\infty} dr = -\pi\rho r^2(\Omega\nu)^{\frac{1}{2}} H_\infty. \qquad (3.26)$$

The frictional moment on one side of a disc of radius b is

$$M = -\int_0^b 2\pi r^2\mu\left[\frac{\partial v_\varphi}{\partial z}\right]_{z=0} dr = -\tfrac{1}{2}\pi\rho\Omega^{\frac{3}{2}}\nu^{\frac{1}{2}}b^4 G'(0). \qquad (3.27)$$

The nondimensional local mass flow-rate for the entrained

flow on the disc is

$$\frac{\dot{m}_0}{\mu r} = -\pi H_\infty (x^2 Re_\varphi)^{\frac{1}{2}} = 2.779(x^2 Re_\varphi)^{\frac{1}{2}}, \qquad (3.28)$$

and the shear stress on the disc is given by

$$\frac{\tau_{\varphi,0}}{\rho\Omega^2 b^2} = G'(0)\, x\, Re_\varphi^{-\frac{1}{2}} = -0.6159\, x\, Re_\varphi^{-\frac{1}{2}}; \qquad (3.29)$$

this implies that the stress-flow-rate parameter, K_w, in equation (2.74) is 0.6964. The moment coefficient, C_M, (see list of symbols) is

$$C_M = -\pi G'(0) Re_\varphi^{-\frac{1}{2}} = 1.935\, Re_\varphi^{-\frac{1}{2}}. \qquad (3.30)$$

The graph of C_M as a function Re_φ is shown in Figures 4.2 and 4.4.

3.1.2 Experiment

Gregory, Stuart and Walker (1955) measured velocity profiles in the boundary layer on a perspex disc of 305 mm diameter rotating in air; they used a two-tube "Conrad probe" to measure the yaw angle, and a pitot tube to measure the resultant total pressure. (For simplicity, such probes will be referred to below as "aerodynamic probes", even if the fluid is not air.) Figure 3.3 shows a comparison of their results for $x^2 Re_\varphi = 1.13 \times 10^5$ and 1.37×10^5 with the theory described above. The variation of the nondimensional resultant velocity

$$\frac{v_T}{\Omega r} = \frac{(v_r^2 + v_\varphi^2)^{\frac{1}{2}}}{\Omega r} \qquad (3.31)$$

with ζ is shown in Figure 3.3a; the agreement between theory and experiment is very good. The agreement is less satisfactory for the angle $\psi = \tan^{-1}(v_r/v_\varphi)$ between the direction of flow and the radius vector (see Figure 3.3b); this was probably due to the experimental errors arising from the inaccuracy of the yawmeter.

The results of a similar experiment by Cham and Head (1969) with $x^2 Re_\varphi = 1.29 \times 10^5$ are also shown in Figure

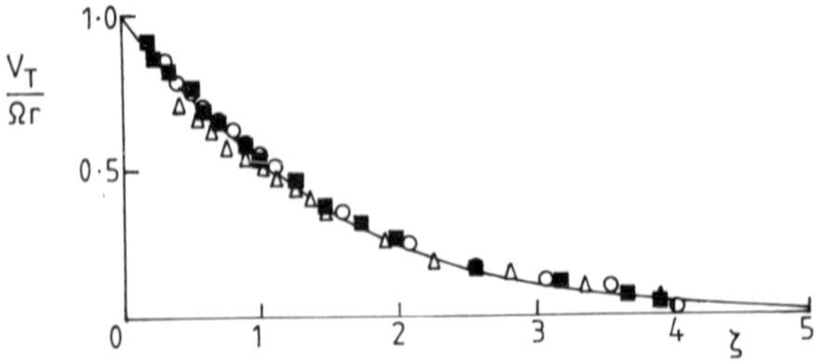

(a)

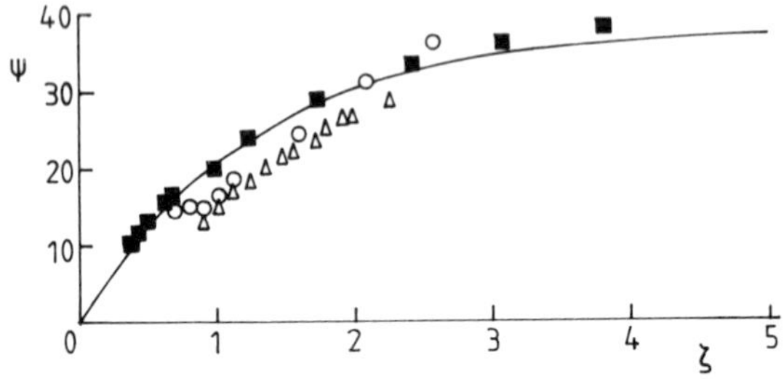

(b)

Figure 3.3 The free disc: variation with ζ of
(a) the nondimensional total velocity, $v_T/\Omega r$
(b) the angle, ψ°, between the direction of flow and the radius vector

Theory:	————		
Experiment:	Δ	$Re_{\phi} = 1.13\text{x}10^5$	Gregory *et al* (1955)
	○	$Re_{\phi} = 1.37\text{x}10^5$	
	■	$Re_{\phi} = 1.29\text{x}10^5$	Cham and Head (1969)

3.3. They used a steel disc of 914 mm diameter rotating in air; the angle ψ was measured by a hot-wire anemometer, and a total-head probe was used to determine v_T. For $\zeta < 2$, their results are in excellent agreement with the theory of Section 3.1.1; for larger values of ζ, the small values of velocity made accurate measurements difficult.

The earliest measurements of C_M were carried out by Kempf (1924) whose values were much higher than the theoretical curve; most authors now consider that his results are inaccurate. The variation of C_M with Re_ϕ given by equation (3.30) is compared in Figure 4.4 with the experimental results of Theodorsen and Regier (1944) who used a disc rotating in either air or Freon 12 to achieve values of Re_ϕ up to 7×10^6. Although they gave no details about how the moments were measured, their data for $5\text{x}10^4 < Re_\phi < 3\text{x}10^5$ were in excellent agreement with equation (3.30). Mach numbers up to approximately 2 were achieved, using Freon 12, but there was no measurable Mach number effect.

3.2 GENERALIZATION OF THE LAMINAR-FLOW EQUATIONS

Equations (3.05) to (3.08) can be generalized in circumstances where it is permissible to write

$$v_r = rAF(\zeta), \quad v_\phi = rAG(\zeta), \quad v_z = (\nu A)^{\frac{1}{2}}H(\zeta) \qquad (3.32)$$

and

$$p = -\rho\nu AP(\zeta) + \tfrac{1}{2}\rho Br^2, \qquad (3.33)$$

where

$$\zeta = (A/\nu)^{\frac{1}{2}}z \qquad (3.34)$$

and A, B are constants. Substitution of these equations into equations (2.02) to (2.06) gives

$$H' = -2F, \qquad (3.35)$$

$$F'' = F^2 - G^2 + F'H + B/A^2 \qquad (3.36)$$

$$G'' = 2FG + G'H, \qquad (3.37)$$

$$P' = HH' - H'', \qquad (3.38)$$

where primes indicate differentiation with respect to ζ. It may be noted that if $A = \Omega$ and $B = 0$, these equations reduce to equations (3.05) to (3.08) for the free disc. In general, however, the values of A and B for the boundary conditions for equations (3.35) to (3.37) depend on the problem under discussion. The solutions may be found, with only slight modifications, using any of the methods described in Section 3.1.

3.2.1 A rotating disc in a rotating fluid

When a disc is rotating with angular speed Ω in a fluid rotating with speed ω about the same axis of rotation, it is necessary to take

$$A = \Omega,\ B = \omega^2 = \beta^2\Omega^2, \tag{3.39}$$

together with the boundary conditions

$$F(0) = 0,\ G(0) = 1,\ H(0) = 0, \tag{3.40}$$

$$F(\zeta) \to 0,\ G(\zeta) \to \beta \quad \text{when } \zeta \to \infty. \tag{3.41}$$

When $\omega = 0$, the system reduces to the free-disc problem of von Kármán (discussed in Section 3.1 above); when $\omega = \Omega$, there is solid-body rotation; and when $\Omega = 0$ (that is, when $\beta \to \infty$), the problem is the same as that discussed by Bödewadt (see Section 3.2.2 below). Solutions to this system of equations have been computed by Rogers and Lance (1960) for values of β between -0.2 and infinity; they were unable to find solutions for $\beta < -0.2$. For application to rotor-stator systems, it is of particular interest to discuss the case for $\beta \geq 0$, and only the results for this case are presented here. When $\beta > 1$, the equations are more easily solved by defining a new independent variable $\xi = \beta^{\frac{1}{2}}\zeta$ and new dependent variables $\tilde{F}(\xi) = \beta^{-1}F(\zeta)$, $\tilde{G}(\xi) = \beta^{-1}G(\zeta)$ and $\tilde{H}(\xi) = \beta^{-\frac{1}{2}}H(\zeta)$.

Computed values of H_∞, $F'(0)$ and $G'(0)$ are given in Table 3.2. The authors claim greater accuracy for their solutions than that given in the tables; however, their value of H_∞ when $\beta = 0$ differs in the fifth decimal place from the accurate value given in equation (3.23), and their value of H_∞ for $\beta \to \infty$ differs from that given in equation (3.67). They stated that they found values of $F'(0)$ and $G'(0)$ to an accuracy of 10 decimal places, and that the asymptotic solution was valid for $\zeta > 6$; more recent computations, on a VAX 8530 computer with a typical precision of 16 significant figures, for the case in which

Table 3.2 Starting and asymptotic values from the numerical solution for a rotating disc in a rotating fluid

(a) $\beta \leq 1$

β	$F'(0)$	$G'(0)$	H_∞	λ_1	λ_2	K_ω
0.0	0.510233	-0.615922	-0.8845	-0.8845	0	0.6964
0.1	0.513397	-0.601554	-0.9177	-0.9593	-0.1998	0.6555
0.2	0.501870	-0.572080	-0.8618	-0.9906	-0.3573	0.6639
0.4	0.439580	-0.478673	-0.6600	-1.0068	-0.5910	0.7253
0.6	0.331470	-0.348434	-0.4308	-1.0051	-0.7598	0.8089
0.8	0.183469	-0.187591	-0.2080	-1.0015	-0.8914	0.9018
0.9	0.095936	-0.096951	-0.1020	-1.0004	-0.9480	0.9504
1.0	0	0	0	-1	-1	1

(b) $\beta \geq 1$

β^{-1}	$\beta^{-\frac{3}{2}}F'(0)$	$\beta^{-\frac{3}{2}}G'(0)$	$\beta^{-\frac{1}{2}}H_\infty$	$\beta^{-\frac{1}{2}}\lambda_1$	$\beta^{-\frac{1}{2}}\lambda_2$	$\beta^{-1}K_\omega$
0.0	-0.941971	0.772886	1.3696	-0.4384	-0.8903	0.5643
0.1	-0.844923	0.718393	1.1952	-0.4953	-0.9150	0.6011
0.2	-0.751682	0.658418	1.0308	-0.5531	-0.9359	0.6387
0.4	-0.569080	0.522477	0.7261	-0.6704	-0.9676	0.7196
0.6	-0.385914	0.366431	0.4538	-0.7861	-0.9872	0.8075
0.8	-0.196121	0.191812	0.2127	-0.8965	-0.9972	0.9017
0.9	-0.099014	0.097977	0.1031	-0.9491	-0.9993	0.9504
1.0	0	0	0	-1	-1	1

$\beta \to \infty$, suggest that it is necessary to specify $F'(0)$ and $G'(0)$ to at least twelve decimal places and that the asymptotic solution discussed above is only valid for $\zeta > 10$. In spite of the doubt which this throws on the accuracy claimed by Rogers and Lance, their solutions are approximately correct and may be used to indicate trends of H_∞, λ_1 and λ_2 as the value of β varies.

Profiles of $F(\zeta)$, $G(\zeta)$ and $H(\zeta)$ for $\omega/\Omega = 0.2$ and 0.8 are shown in Figure 3.4 for $\beta < 1$ and in Figure 3.5 (see Section 3.2.2) for $\beta \to \infty$. It can be seen that, for $\beta = 0.8$, there is an oscillatory behaviour as $\zeta \to \infty$; the spacing of the zeros is approximately equal to π. For $\beta = 0.2$ the oscillations are of smaller magnitude and the spacing is larger; in this case the profiles approach those of the free-disc solution already discussed. For $\beta > 1$, the oscillations are much larger in magnitude.

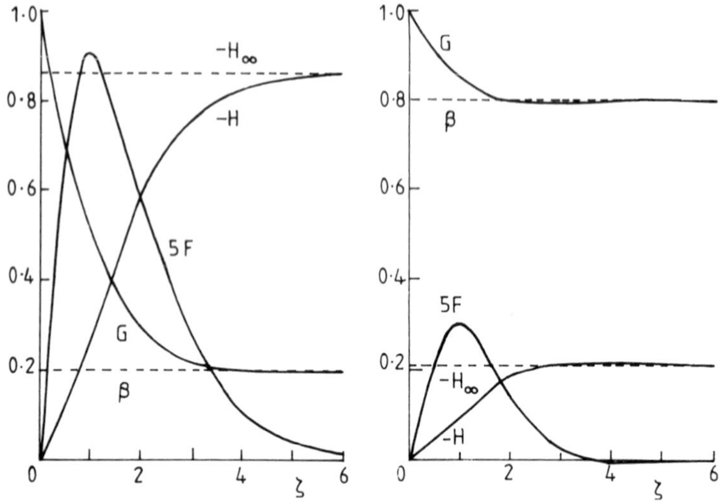

(a) $\beta = 0.2$ (b) $\beta = 0.8$

Figure 3.4 Velocity profiles for a rotating disc in a rotating fluid (Rogers and Lance 1960)

The nondimensional mass flow-rate, the stress-flow-rate parameter (see equation (2.74)) and the moment coefficient are given by

$$\frac{\dot{m}_0}{\mu r} = -\pi H_\infty (x^2 Re_\phi)^{\frac{1}{2}}, \quad K_w = \frac{G'(0)}{H_\infty}, \quad C_M = -\pi G'(0) Re_\phi^{-\frac{1}{2}}, \quad (3.42)$$

where values for H_∞, $G'(0)$ and K_w are given in Table 3.2. (When $\beta = 0$, these values reduce to those found in Section 3.1.1 for the free disc.)

The behaviour of the solution as $\zeta \to \infty$ is of particular interest, since the approach to the asymptotic value is oscillatory. In particular, as ζ increases, there are alternate layers of radial inflow and outflow. An understanding of how this comes about may be obtained by considering the motion of a fluid element which, for $\omega/\Omega < 1$, starts far from the disc (in the "core") and moves towards the disc; at the same time the element begins to move radially outwards. In the core, viscous forces are negligible and there is a balance between centrifugal and pressure-gradient forces. When there is a radial displacement of the fluid element, its angular momentum is conserved and this causes an imbalance of forces. The effect of this imbalance is to tend to return the element to its original radial position and, in overshooting, an oscillation is set up. As the element approaches the disc, it must move further from the axis; the magnitude of the oscillation therefore increases until the general radial motion overrides the oscillatory effect. For a real fluid, viscosity has the effect of damping the oscillations (but not sufficiently to eliminate them) and of forcing the velocity of the fluid relative to the disc to be zero at the surface. Associated with the radial oscillations, there are corresponding oscillations of the tangential and axial components of velocity, but these (unlike the radial component) do not undergo a change of sign.

The same argument can be used for the case in which $\beta > 1$. In this case, the fluid element moves away from the disc and its radial component of velocity is initially large. Oscillations, once they start, are also large and, if viscosity is neglected, there is no reason for them to decrease in magnitude. However, for a real fluid, viscous effects cause a diminution of the magnitude of the oscillations so that they die away far from the disc. It is not surprising, therefore, that the oscillations are of larger magnitude and extend further from the disc than when the flow is towards the disc.

The asymptotic flow as $\zeta \to \infty$ is investigated by writing

$$F(\zeta) = f(\zeta), \; G(\zeta) = \beta + g(\zeta), \; H(\zeta) = -c + h(\zeta), \qquad (3.43)$$

where $c = -H_\infty$ (the minus sign is conventionally used here for consistency with the free-disc problem); it is assumed that, when ζ is large, f, g, h and their derivatives are small. Then, substituting equations (3.43) into equations (3.36) and (3.37), and neglecting squares and products of small quantities, it is found that

$$f'' + cf' + 2\beta g = 0, \qquad g'' + cg' - 2\beta f = 0. \tag{3.44}$$

These are linear differential equations with constant coefficients and, remembering that f and g approach zero as $\zeta \to \infty$, their solution may easily be found. The easiest way to do this is to define the complex function

$$q(\zeta) = f(\zeta) + ig(\zeta). \tag{3.45}$$

Equations (3.44) then combine to give

$$q'' + cq' - 2i\beta q = 0. \tag{3.46}$$

The solution to this equation is of the form

$$q(\zeta) = C_0 \exp[(\lambda_1 + i\lambda_2)(\zeta - \zeta_0)], \tag{3.47}$$

where C_0, λ_1, λ_2 and ζ_0 are real constants; to satisfy the conditions as $\zeta \to \infty$, it is necessary that $\lambda_1 < 0$. Substituting from equation (3.47) into equations (3.44) and equating real and imaginary parts gives

$$\lambda_1{}^2 - \lambda_2{}^2 + c\lambda_1 = 0, \; 2\lambda_1\lambda_2 + c\lambda_2 - 2\beta = 0. \tag{3.48}$$

The second of these equations gives λ_2 in terms of λ_1, and this can be used in the first equation; it follows that

$$(\lambda_1 + \tfrac{1}{2}c)^4 - \tfrac{1}{4}c^2(\lambda_1 + \tfrac{1}{2}c)^2 - \beta^2 = 0. \tag{3.49}$$

This equation is easily solved to give

$$\lambda_1 = -\frac{1}{2}\left[c + \left(\frac{1}{2}[c^2 + (c^4 + 64\beta^2)^{\frac{1}{2}}]\right)^{\frac{1}{2}}\right], \tag{3.50}$$

and

$$\lambda_2 = \beta/(\lambda_1 + \tfrac{1}{2}c). \tag{3.51}$$

Then, reverting to the original variables $F(\zeta)$ and $G(\zeta)$, it may be seen that, for large ζ,

$$F(\zeta) \simeq C_0\, e^{\lambda_1(\zeta-\zeta_0)} \cos[\lambda_2(\zeta - \zeta_0)], \tag{3.52}$$

$$G(\zeta) \simeq -\beta - C_0\, e^{\lambda_1(\zeta-\zeta_0)} \sin[\lambda_2(\zeta - \zeta_0)] \tag{3.53}$$

where C_0 and ζ_0 are constants which must be determined by matching the numerical solution to the full equations. Values of λ_1 and λ_2 are shown in Table 3.2. The sign of c cannot be deduced from the asymptotic solution (as it was for the free disc), but it is always positive when $0 < \beta < 1$ and negative when $\beta > 1$. It is also found that, as β increases from zero to infinity, $|\lambda_1|$ increases and $\pi/|\lambda_2|$ decreases: this means that the rate of decay of the oscillations and the spacing of the zeros of the solution decrease as β increases.

Although it is not expected that the approximations made to obtain the Ekman-layer equations (see Section 2.2.2) are strictly valid except when β is near unity, it is useful to investigate the solution of these equations. The advantage of the approximations is that they give a solution which (with one exception) is qualitatively similar to the exact solution and which differs from that solution by less than an order of magnitude.

Since the flow is laminar, equations (2.44) and (2.45) may be written in the form

$$-2\Omega[v - (\omega - \Omega)r] = \nu\frac{\partial^2 u}{\partial z^2}, \qquad 2\Omega u = \nu\frac{\partial^2 v}{\partial z^2}, \tag{3.54}$$

and the boundary conditions are

$$u(0) = v(0) = 0, \tag{3.55}$$

$$u(\zeta) \to 0,\ v(\zeta) \to -(\Omega - \omega)r \quad \text{as } \zeta \to \infty. \tag{3.56}$$

To solve these equations it is convenient to define the complex function

$$q(\zeta) = u + i[v - (\omega - \Omega)r]. \tag{3.57}$$

Then q satisfies the equation

$$\frac{\partial^2 q}{\partial \zeta^2} = (1 + i)^2 q \tag{3.58}$$

and the boundary conditions

$$q(0) = i(\Omega - \omega)r, \quad q \to 0 \quad \text{as } \zeta \to \infty. \tag{3.59}$$

The solution is therefore

$$q = -i(\omega - \Omega)r \exp[-(1 + i)\zeta]. \tag{3.60}$$

and the real components of relative velocity are given by

$$u = (\Omega - \omega)re^{-\zeta}\sin\zeta, \quad v = -(\Omega - \omega)r(1 - e^{-\zeta}\cos\zeta). \tag{3.61}$$

The approximate value of w may be calculated using the equation of continuity: it is found that

$$w = -\left(\frac{\nu}{\Omega}\right)^{\frac{1}{2}}(\Omega - \omega)[1 - e^{-\zeta}(\cos\zeta + \sin\zeta)]. \tag{3.62}$$

In order to compare this solution with the numerical solution of the full equations, it may be seen that it gives

$$H_\infty \simeq -(1 - \beta), \ \lambda_1 \simeq -1, \ \lambda_2 \simeq -1 \tag{3.63}$$

and
$$\frac{\dot{m}_0}{\mu r} = \pi(1 - \beta)(x^2 Re_\phi)^{\frac{1}{2}}. \tag{3.64}$$

The relationship (2.74) is valid with $K_w = 1$.

It may be seen from Table 3.2 that, in the range $0.1 \leq \beta \leq 1.25$, $|H_\infty|$ is in error by less than 10% (and H_∞ is of the correct sign), that λ_1 is in error by less than 5%, but that the values of λ_2 and K_w are considerably in error (except for $0.9 < \beta < 1.11$). When $\beta = 1$, the approximate solution gives solid-body rotation, as it should. When $\beta = 0$, however, the method gives a spurious oscillatory behaviour at large values of ζ: nevertheless, even in this case, the approximate values of H_∞ and λ_1 are less than 16% in error.

Experimental evidence to support the theory of this

section has mostly been carried out in an enclosed rotor-stator system. Discussion of this will, therefore, be postponed to Chapters 6 and 7.

3.2.2 A stationary disc in a rotating fluid

For a stationary disc in a fluid rotating with angular speed ω, $A = \omega$, $B = \omega^2$ so that equations (3.35) to (3.37) are identical with those of the previous section when $B/A^2 = 1$. The boundary conditions are

$$F(0) = 0,\ G(0) = 0,\ H(0) = 0, \tag{3.65}$$

$$F(\zeta) \to 0,\ G(\zeta) \to 1 \quad \text{as } \zeta \to \infty. \tag{3.66}$$

This case was first discussed by Bödewadt (1940) who solved the equations by a method similar to that used by Cochran for the von Kármán equations. The results have also been computed by Rogers and Lance (1960) and, more recently, by Nydahl (see Schlichting 1979). Nydahl's method is not specified but his results effectively confirm those of Bödewadt; those of Rogers and Lance give, as stated in Section 3.2.1, a significantly larger value of H_∞.

Computations have recently been carried out (using the method of Rogers and Lance and an integrating routine given by Shampine and Gordon 1975) on a VAX 8530 computer with a typical precision of 16 significant figures. These computations (which suggest that the accuracy achieved by Rogers and Lance was not sufficient to give an accurate value of H_∞) are presented in Table 3.3 and Figure 3.5; they are in agreement with virtually all the results of Nydahl. The numerical solution gives

$$H_\infty = 1.3494,\ F'(0) = -0.94197,\ G'(0) = 0.77289, \tag{3.67}$$

and these values, together with the expressions in equation (3.42), give

Table 3.3 Velocity profiles for a stationary disc in a rotating fluid

ζ	F	G	H
0.0	0.0000	0.0000	0.0000
0.5	-0.3487	0.3834	0.1944
1.0	-0.4788	0.7354	0.6241
1.5	-0.4496	1.0134	1.0987
2.0	-0.3287	1.1924	1.4929
2.5	-0.1762	1.2721	1.7459
3.0	-0.0361	1.2714	1.8496
3.5	0.0663	1.2182	1.8308
4.0	0.1227	1.1413	1.7325
4.5	0.1371	1.0640	1.5995
5.0	0.1210	1.0016	1.4685
5.5	0.0878	0.9611	1.3632
6.0	0.0499	0.9427	1.2944
6.5	0.0162	0.9421	1.2620
7.0	-0.0084	0.9530	1.2589
7.5	-0.0223	0.9693	1.2751
8.0	-0.0268	0.9857	1.3004
8.5	-0.0243	0.9990	1.3264
9.0	-0.0179	1.0078	1.3476
9.5	-0.0102	1.0118	1.3617
10.0	-0.0033	1.0121	1.3683
10.5	0.0018	1.0099	1.3689
11.0	0.0047	1.0065	1.3654
11.5	0.0057	1.0031	1.3601
12.0	0.0052	1.0003	1.3545
12.5	0.0038	0.9984	1.3500
∞	0.0000	1.0000	1.3494

$$\frac{\dot{m}_0}{\mu r} = -4.239\,(x^2 Re'_\phi)^{\frac{1}{2}}, \tag{3.68}$$

where

$$Re'_\phi = \frac{\rho\omega b^2}{\mu} \tag{3.69}$$

is a modified rotational Reynolds number related to the angular speed of the fluid. The shear stress is related to the local mass flow-rate and the rotational Reynolds number by the relation

$$\frac{\tau_{\phi,0}}{\rho\omega^2 b^2} = -\frac{1}{\pi} K'_w \, Re'^{-1}_\phi \frac{\dot{m}_0}{\mu r}, \tag{3.70}$$

where $K'_w = 0.6167$ is analogous to the stress-mass-flow parameter K_w which occurs in equation (2.74).

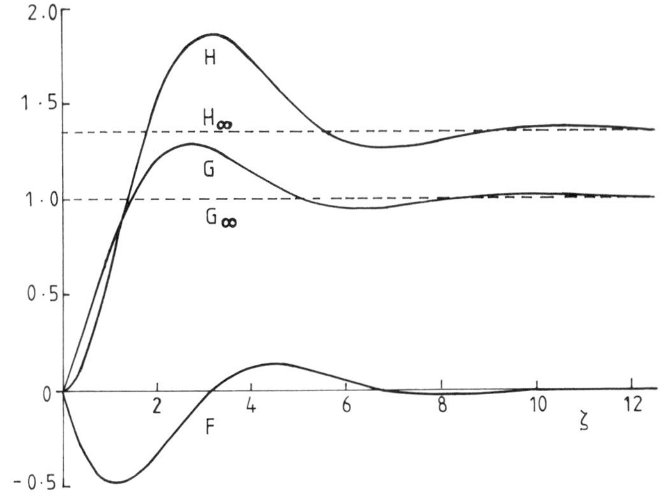

Figure 3.5 Velocity profiles for a stationary disc in a rotating fluid (Bödewadt 1940)

The flow is, of course, the same as that discussed in Section 3.2.1 when $\beta \to \infty$; it is radially inwards and away from the disc, and the oscillations are large. (It may be of interest to note that this radially inward flow is the cause of the accumulation of sugar at the centre of a stirred cup of tea!)

An asymptotic solution may be found, in the same way as that used for a rotating disc in a rotating fluid. With $\alpha = 1$ and H_∞ given by the first of equations (3.67), it is easy to show that

$$\lambda_1 = -0.4447, \qquad \lambda_2 = -0.8933; \tag{3.71}$$

and equations (3.52) and (3.53), with $\beta = 1$, are still valid.

If the Ekman-layer approximation is made with $\Omega' = \omega$, it is easily shown that

$$u = -\omega r e^{-\zeta} \sin\zeta, \quad v = -\omega r e^{-\zeta} \cos\zeta, \tag{3.72}$$

$$w = (\nu\omega)^{\frac{1}{2}} [1 - e^{-\zeta}(\cos\zeta + \sin\zeta)]. \tag{3.73}$$

This gives

$$H_\infty \simeq 1, \ \lambda_1 \simeq -1, \ \lambda_2 \simeq -1, \ A = 0, \ B \simeq -1 \qquad (3.74)$$

and

$$\frac{\dot{m}_0}{\mu r} = \pi(x^2 Re'_\phi)^{\frac{1}{2}}; \qquad (3.75)$$

it follows that H_∞ is underestimated by about 35%, the asymptotic decay is faster by a factor of just over 2 and the spacing of the zeros of u is too small by about 10%. The method gives, once again, a good qualitative picture of the flow; quantitatively, the errors are less than an order of magnitude.

It has been pointed out by many authors that Bödewadt's solution implies that there is flow out of the boundary layer everywhere and no mechanism for supplying fluid to it! For an infinite disc, the problem may be overcome by assuming an infinite reservoir of fluid from which the boundary layer can draw in an unspecified way. For the more practical case of a finite disc, it must be supposed that a similarity solution does not hold near the edge of the disc. However, it is consistent with experience in other fields that a similarity solution becomes valid as the boundary-layer flow develops. It may then be assumed that fluid enters the boundary layer near the edge of the disc and that this fluid is available for continuity in the similarity solution. (For a shrouded rotor-stator system, fluid is transferred to the stator from the rotor via a boundary layer on the shroud: this is illustrated in Section 7.3.2.)

As in the previous section, comparison with experimental data is postponed to Chapters 6 and 7.

3.2.3 Flow impinging on a rotating disc

Milne-Thomson (1960) discussed the irrotational flow when a flat disc of radius b is moving with a speed U normal to its surface; his solution shows that the radial component of velocity on the disc, $v_{r,\infty}$, is given by

$$v_{r,\infty} = \frac{2U}{\pi} \frac{r}{(b^2 - r^2)^{\frac{1}{2}}} . \tag{3.76}$$

This value is changed neither by the superposition of a uniform stream on the flow (so that there is an impinging flow on a stationary plate) nor by a rotation of the disc with angular speed Ω about an axis parallel to the flow. It may, therefore, be used as an outer boundary condition for the boundary layer on the disc in such a flow. If both U and b are known, it is convenient to write

$$\sigma = \frac{2}{\pi} \frac{U}{\Omega b} ; \tag{3.77}$$

then, as long as $r \ll b$, equation (3.76) becomes, to a good approximation, $v_{r,\infty} = \sigma\Omega r$.

Tifford and Chu (1954) showed that equations (3.35) to (3.37) are valid with $A = (\sigma + 1)\Omega$, and $B = -\sigma^2\Omega^2$; they may therefore be written as

$$H' = -2F \tag{3.78}$$

$$F'' = F^2 - G^2 + F'H - \frac{\sigma^2}{(\sigma + 1)^2} \tag{3.79}$$

$$G'' = 2FG + G'H \tag{3.80}$$

The boundary conditions are

$$F(0) = 0, \quad G(0) = \frac{1}{\sigma + 1}, \quad H(0) = 0, \tag{3.81}$$

$$F(\zeta) \rightarrow \frac{\sigma}{\sigma + 1}, \quad G(\zeta) \rightarrow 0 \quad \text{as } \zeta \rightarrow \infty. \tag{3.82}$$

The equations were solved using a method similar to that used by Cochran for equations (3.04) to (3.08) for various values of σ. From their solutions, they computed the local moment coefficient as a function of σ, and their results are presented in Table 3.4. Since the solution does not hold for the whole disc, and the integration to determine C_M is carried out from $r = 0$ to b, the results must be treated with caution.

Table 3.4 Values of the moment coefficient for impinging flow on a rotating disc (Tifford and Chu 1954)

σ	0	0.1	0.25	0.5	1.0	1.5	2.0	4.0	6.0
$C_M Re_\phi^{\frac{1}{2}}$	1.94	2.03	2.24	2.65	3.21	4.20	4.82	6.67	8.32

CHAPTER 4
Turbulent Flow over a Single Disc

4.1 TRANSITION FROM LAMINAR TO TURBULENT FLOW ON THE FREE DISC

As with all boundary-layer flows, there are two stages in the transition from laminar to turbulent flow. For low values of the local Reynolds number, $x^2 Re_\phi$, the flow is laminar; this laminar flow becomes unstable at a certain value of $x^2 Re_\phi$, and it is to be expected that the resulting flow either contains small patches of turbulence or consists of a more regular disturbance. This "disturbed laminar flow" finally breaks down at a critical value of $x^2 Re_\phi$ and the flow becomes fully turbulent.

Smith (1947) was the first to observe regular oscillations when the laminar flow became unstable, but the first major investigation, both theoretical and experimental, was by Gregory, Stuart and Walker (1955). Their work will be described in some detail here.

Stuart (see Gregory *et al*) found a solution that bears some relation to the experiments in which "disturbed laminar flow" is observed. He looked for small three-dimensional perturbations to Cochran's (1934) solution of von Kårmån's (1921) equations (see Section 3.1.1) and obtained an Orr-Sommerfeld equation (see Schlichting 1979). Neglecting the viscous terms in his equations, he used a variational technique to find a steady solution (which was not unique) consisting of *n* small vortices in the form of logarithmic spirals making a constant angle ψ with the inward-pointing radius vector. Stuart found that $\psi = 76.7°$ (which is in good agreement with experiment) and n had a value between 113 and 140 (which is much too large). The solution is shown schematically in Figure 4.1 in which each vortex is represented by a single line (which may be thought of as the axis of the vortex); the value of ψ is as predicted by Stuart, but only 31 vortices are shown. Since the theory was essentially an inviscid one, Stuart could give no range of values of $x^2 Re_\varphi$ for which his solution was valid.

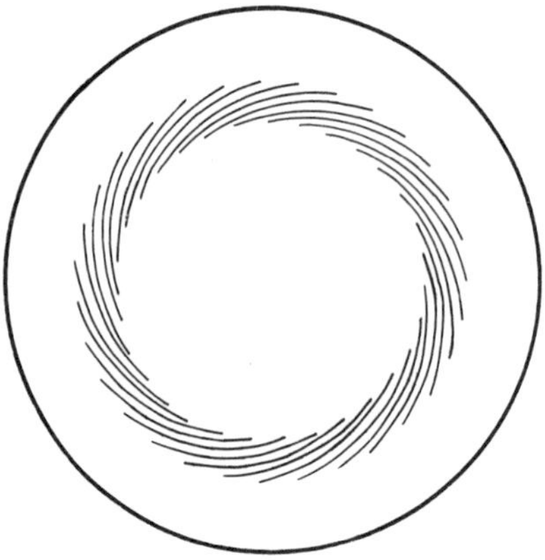

Figure 4.1
Schematic diagram of vortices occurring when laminar flow becomes unstable

Brown (1961) considered the same equations as Stuart, but retained the viscous terms; he looked for a closed analytic solution outside the boundary layer and matched this, at a suitable distance from the disc, to a numerical solution within the boundary layer. He obtained stability curves which indicate that there can be no perturbations of the laminar solution for $x^2 Re_\varphi < 3.2 \times 10^4$.

Gregory *et al* (1955) used a china-clay (kaolin) technique for the visualization of the flow on the surface of a Perspex disc. The disc was of diameter 305 mm and rotated in air at angular speeds between 600 and 3200 rev/min. They also used an acoustical stethoscope, held close to the surface of the disc, to differentiate between "silence" in the laminar region, a "regular note" in the transition region and a "roar" in the turbulent region. As a result of their experiments, they found that laminar flow became unstable when $x^2 Re_\phi$ was between 1.78×10^5 and 2.12×10^5 and that transition to turbulence was complete for $x^2 Re_\phi$ between 2.70×10^5 and 2.99×10^5; they attributed the scatter in these results to turbulence in the surrounding air.

Their china-clay experiments also showed a clear indication of the presence of the vortex system described by Stuart. The observed pattern was made up of several vortex systems, the predominant system being stationary relative to the disc. For this stationary system, the measured angle ψ was 76°, and this agrees very well with Stuart's predicted value of 76.7°. However, the number of vortices present was between 28 and 31, a value very different from the theoretical prediction; Stuart suggested that the discrepancy was due to the neglect of viscosity in his solution and this is confirmed by the theoretical work of Brown.

More recent theoretical work by Faller and Kaylor (1966a), Lilly (1966), Tobak (1973), Yamashita and Takematsu (1974), Kobayashi, Kohama and Takamadate (1980), Cebeci and Stewartson (1980) and Malik, Wilkinson and Orszag (1981) have shown that the inclusion of Coriolis acceleration and streamline curvature give secondary instabilities, some of which occur at much lower values of $x^2 Re_\phi$.

In their experiments, Theodorsen and Regier (1944) found that departure from laminar flow on a highly polished disc began at about $x^2 Re_\phi = 3.1 \times 10^5$ (on a roughened disc, they found that this value could be as low as 2.2×10^5), and fully-established turbulent flow occurred at $x^2 Re_\phi \simeq 7\times10^5$. In the transition region the authors, using hot-wire techniques, found a pure tone output whose frequency was about 200Hz; this, presumably, is related to the steady perturbation found theoretically by Stuart.

More recent experiments by Kreith, Taylor and Chong (1959), Gregory and Walker (1960), Faller and Kâylor (1966b), Chin and Litt (1972a,b), Fedorov, Plavnik, Prokhorov and Zhukhovitskii (1976), Kobayashi, Kohama and Takamadate (1980), Clarkson, Chin and Shacter (1980), Malik, Wilkinson and Orszag (1981) and Wilkinson and Malik (1985) have supported the theoretical predictions very well. A detailed summary of the work up to 1981 is given by Malik *et al* (1981).

4.2 TURBULENT FLOW OVER THE FREE DISC

The structure of turbulent flow over a rotating disc is the same as that for the laminar flow, and the boundary conditions given in equation (3.01) also apply. There are, however, no explicit expressions for the turbulent shear stresses in terms of the velocity and its derivatives. It is convenient to summarize below the well-established approaches used for flow over a smooth flat plate and through a pipe (see Schlichting 1979).

4.2.1 Turbulent shear stress for a stationary surface

For turbulent flow in stationary systems, it is common to write

$$\frac{u}{v^*} = A \ln\left(\frac{\rho v^* z}{\mu}\right) + B, \tag{4.01}$$

where u is the component of velocity parallel to the plate or along the pipe, z is the normal distance from the surface, and v^* is the *friction velocity* defined by

$$v^* = \left(\frac{\tau_0}{\rho}\right)^{\frac{1}{2}}, \tag{4.02}$$

(τ_0 being the shear stress at the surface) and A, B are constants often taken to be $A = 2.5$ and $B = 5.5$. This result may be derived using Prandtl's mixing-length theory, taking the mixing length proportional to z and assuming that the shear stress is independent of z. The theoretical result using

$$A = 2.46, \quad B = 5.69 \tag{4.03}$$

agrees well with pipe-flow experiments carried out by Nikuradse (1932) for values of $\rho v^* z/\mu$ greater than about 70. Nearer the surface than this, viscous forces are important and the logarithmic law given in equation (4.01) no longer holds; the region close to the surface is usually referred to as the *viscous sublayer*.

It may easily be demonstrated that, for considerable ranges of $\rho v^* z/\mu$, the logarithmic expression on the right-hand side of equation (4.01) can be approximated very accurately by a *power law*. The equation may then be written as

$$\frac{u}{v^*} = C\left(\frac{\rho v^* z}{\mu}\right)^{\frac{1}{n}}, \tag{4.04}$$

where the coefficient C is given in Table 4.1 for four values of n. These values of C are those given by

---

Table 4.1 Values of constants for power-law profiles

Coeff-icient	Equation number	n=7	n=8	n=9	n=10
C	(4.04)	8.74	9.71	10.6	11.5
K	(4.06)	0.0225	0.0176	0.0143	0.0118
α	(4.19)	0.162	0.143	0.128	0.116
K_H	(4.23)	0.383	0.348	0.318	0.273
γ	(4.24)	0.526	0.497	0.479	0.463
ϵ_m	(4.25)	0.219	0.187	0.164	0.146
ϵ_H	(4.26)	0.0729	0.0561	0.0448	0.0365

Wieghardt (1946) and are good approximations for different ranges of Reynolds number. As long as n is chosen appropriately, the expression on the right-hand side of equation (4.04) differs from that of equation (4.01) by less than 1%. The case for which $n = 7$ (the $\frac{1}{7}$-*power-law-profile*) is used extensively. (As presented here, the power laws of equation (4.04) are derived as an approximation to the universal logarithmic law. Historically, the $\frac{1}{7}$-power-law profile was discovered by Prandtl 1921 from the study of experimental results.)

Although the range of application of a given power law is limited, it has the advantage of giving an explicit expression for the shear stress at the surface. Rearrangement of equation (4.04) gives, using equation (4.02),

$$\frac{\tau_0}{\rho U^2} = K\left(\frac{\nu}{U r_0}\right)^{\frac{2}{n+1}}, \qquad (4.05)$$

where, for flow in a pipe of radius r_0, U is the value of u on the axis of the pipe and

$$K = C^{-\frac{2n}{n+1}}. \qquad (4.06)$$

The values of K for four values of n are given in Table 4.1. It is of interest to note that, with $K = 0.02335$ and $n = 7$, equation (4.05) is the same as the Blasius (1913) law of friction derived from experiment.

Until recent years, investigation of the turbulent boundary layer on a rotating disc has been dependent on the use of momentum-integral techniques (such as those described in Section 2.4) and has used expressions for the relationships involving the shear stress on the disc which have been established for pipe-flow. First, von Kármán (1921) used power-law distributions similar to equations (4.04) and (4.05); later Goldstein (1935) used a logarithmic profile similar to that of equation (4.01). Both these integral methods are presented here in forms more general than those of the original authors. More recently, Cebeci and Abbott (1975) have found numerical solutions of the partial differential equations (2.26) to (2.28), using a suitable eddy viscosity for the shear stresses; a summary of their work is also given here.

4.2.2 Momentum-integral technique for a rotating disc (using power-law profiles)

The power-law profiles developed for pipe flow can be extended to the flow over a disc of radius b rotating with angular velocity Ω. The detailed discussion in this section is given for a $\frac{1}{7}$-power-law profile (see equations (2.77)) in which it is assumed that, for $\zeta \leq 1$,

$$v_r = \alpha(r)\,\Omega r\, f(\zeta) = \alpha(r)\Omega r\, \zeta^{\frac{1}{7}}(1 - \zeta), \tag{4.07}$$

$$v_\phi = \Omega r\,[1 - g(\zeta)] = \Omega r\,(1 - \zeta^{\frac{1}{7}}), \tag{4.08}$$

where
$$\zeta = \frac{z}{\delta(r)} \tag{4.09}$$

and $\alpha(r)$ is a nondimensional amplitude factor; results are also quoted for a general $\frac{1}{n}$-power law. The method is a direct application of Section 2.4.1 with $v_{\phi,0} = \Omega r$ and $v_{\phi,\infty} = 0$. The extra factor $(1 - \zeta)$ in the profile for v_r is included to ensure that the velocity is zero far from the disc. Equations (4.07) and (4.08) are the same profiles as those given in equations (2.77).

To find expressions for $\tau_{r,0}$ and $\tau_{\phi,0}$, relationships similar to those for pipe flow are assumed, with u replaced by

$$\left[(v_\phi - v_{\phi,0})^2 + v_r^2\right]^{\frac{1}{2}} \simeq v_{\phi,0}(1 + \alpha^2)^{\frac{1}{2}}\zeta^{\frac{1}{7}} \tag{4.10}$$

when $\zeta \ll 1$. This is equivalent to taking $U = \Omega r(1 + \alpha^2)^{\frac{1}{2}}$ in equation (4.05), and r_0 is replaced by δ; then

$$\frac{\tau_0}{\rho\Omega^2 b^2} = K x^2(1 + \alpha^2)\left[\frac{\nu}{\Omega r(1 + \alpha^2)^{\frac{1}{2}}\delta}\right]^{\frac{1}{4}}, \tag{4.11}$$

where $x = r/b$. In addition it is assumed that, as in the laminar case,

$$\frac{\tau_{r,0}}{\tau_{\varphi,0}} = \lim_{z\to 0}\left(\frac{v_r}{v_\varphi - v_{\varphi,0}}\right) = -\lim_{\zeta\to 0}\left(\frac{\alpha f'(\zeta)}{g'(\zeta)}\right) = -\alpha. \qquad (4.12)$$

It follows that $\tau_{r,0}$ and $\tau_{\varphi,0}$ are given by equations (2.78) and (2.79) with, once again, $v_{\varphi,0} = \Omega r$ and $v_{\varphi,c} = 0$.

Equations (2.80) to (2.82) therefore give

$$\dot{m}_0 = \frac{49\pi}{60}\rho\Omega r^2\alpha\delta, \qquad (4.13)$$

$$\frac{343}{1656}\left(3 + 2\frac{r}{\alpha}\frac{d\alpha}{dr} + \frac{r}{\delta}\frac{d\delta}{dr}\right) - \frac{1}{36}\frac{1}{\alpha^2} = -0.0225(x^2 Re_\varphi)^{-\frac{1}{4}}\frac{(1+\alpha^2)^{\frac{3}{8}}}{\alpha}\left(\frac{\delta}{r}\right)^{-\frac{5}{4}}, \qquad (4.14)$$

$$\frac{49}{720}\left(4 + \frac{r}{\alpha}\frac{d\alpha}{dr} + \frac{r}{\delta}\frac{d\delta}{dr}\right) = 0.0225\,(x^2 Re_\varphi)^{-\frac{1}{4}}\frac{(1+\alpha^2)^{\frac{3}{8}}}{\alpha}\left(\frac{\delta}{r}\right)^{-\frac{5}{4}}. \qquad (4.15)$$

If it is assumed, by analogy with the laminar case, that α is independent of r, it is easy to show that

$$\alpha = \left(\frac{2300}{87661}\right)^{\frac{1}{2}} = 0.1620, \quad \frac{\delta}{r} = 0.5261(x^2 Re_\varphi)^{-\frac{1}{5}}. \qquad (4.16)$$

The local nondimensional flow-rate through the boundary layer and the stress-flow-rate parameter (see equation (2.74)) are given by

$$\frac{\dot{m}_0}{\mu r} = 0.2186(x^2 Re_\varphi)^{\frac{4}{5}}, \quad K_w = 0.3833; \qquad (4.17)$$

the nondimensional shear stress and the moment coefficient are

$$\frac{\tau_{\varphi,0}}{\rho\Omega^2 b^2} = -0.02668\,x^{\frac{8}{5}}\,Re_\varphi^{-\frac{1}{5}}, \quad C_M = 0.07288\,Re_\varphi^{-\frac{1}{5}}. \qquad (4.18)$$

For a general $\frac{1}{n}$-power law, it may be shown that

$$\alpha^2 = \frac{4(3n+2)(2n+1)(n+3)}{n^2(16n^3+85n^2+145n+66)}, \tag{4.19}$$

$$\frac{\delta}{r} = \gamma(x^2 Re_\phi)^{-\frac{2}{n+3}}, \tag{4.20}$$

$$\frac{\dot{m}_0}{\mu r} = \epsilon_m (x^2 Re_\phi)^{\frac{n+1}{n+3}}, \tag{4.21}$$

$$C_M = \epsilon_M Re_\phi^{-\frac{2}{n+3}}, \tag{4.22}$$

and

$$K_w = \frac{3(5n + 11)}{4(n + 2)(n + 3)}, \tag{4.23}$$

where

$$\gamma^{\frac{n+3}{n+1}} = K \frac{2(n+1)(n+2)(n+3)(2n+1)(1+\alpha^2)^{(n-1)/2(n+1)}}{3n^2(5n+11)\alpha}, \tag{4.24}$$

$$\epsilon_m = \frac{2n^2 \pi \alpha \gamma}{(n+1)(2n+1)}, \tag{4.25}$$

and

$$\epsilon_M = \frac{6n^2 \pi \alpha \gamma}{(n+1)(n+2)(2n+1)}. \tag{4.26}$$

Some values of α, γ, ϵ_m, ϵ_M and K_w are given in Table 4.1. The variation of C_M with Re_ϕ is shown in Figure 4.2 for $n = 7$ and $n = 10$.

4.2.3 Momentum-integral technique for a rotating disc (using logarithmic profiles)

It is assumed that the profile given in equation (4.01) for pipe flow is also valid for the flow near to a rotating disc, with the velocity u replaced by $[v_r^2 + (v_\phi - \Omega r)^2]^{\frac{1}{2}}$, the magnitude of the velocity relative to the disc. Using the notation of Section 2.4.1 with $v_{\phi,0} = \Omega r$ and $v_{\phi,c} = 0$, and assuming that α is independent of z, it follows that

$$\frac{\Omega r(1 + \alpha^2)^{\frac{1}{2}}}{v^*} = A \ln\left(\frac{\rho v^* \delta}{\mu}\right) + B; \tag{4.27}$$

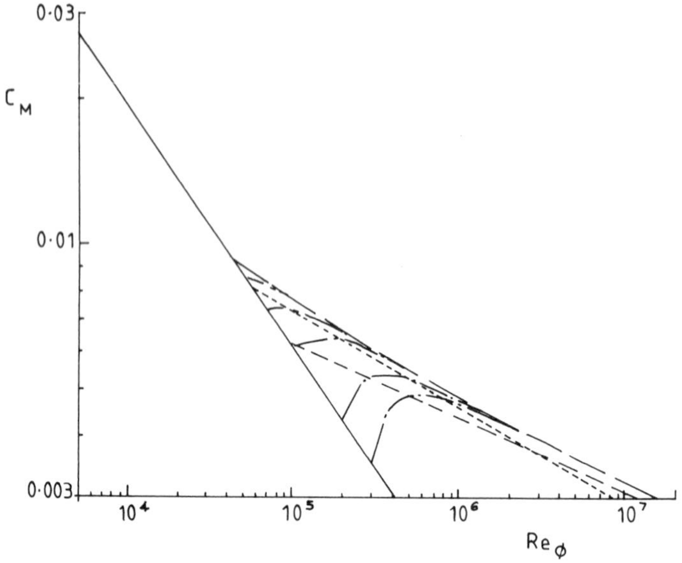

Figure 4.2 Theoretical variation of C_M with Re_φ

————————	laminar flow
- - - - - -	$\frac{1}{7}$-power-law profile
— — — —	$\frac{1}{10}$-power-law profile
—— —— —	asymptotic logarithmic profile
— · — · —	transitional logarithmic profile

---

this is an implicit expression ıor v^* (and hence for τ_0) in terms of r and the dependent variables α and δ. The components of velocity are given by

$$v_\varphi = -\frac{Av^*}{(1 + \alpha^2)^{\frac{1}{2}}} \ln\zeta \qquad (4.28)$$

and

$$v_r = \alpha\Omega r + \frac{A\alpha v^*}{(1 + \alpha^2)^{\frac{1}{2}}} \ln\zeta, \qquad \zeta \leq \zeta_1, \qquad (4.29)$$

where $\zeta = z/\delta$ as before. The limitation on ζ in equation (4.29) is intrıduced to ensure that $v_r \to \infty$ as $\zeta \to 1$. Following Gold: ein (1935), it is assumed that

$$v_r = -\frac{A\alpha v^*}{(1+\alpha^2)^{\frac{1}{2}}}\ln\zeta, \qquad \zeta > \zeta_1 \tag{4.30}$$

where, to ensure that v_r is continuous when $\zeta = \zeta_1$,

$$\zeta_1 = \exp\left(-\frac{\Omega r(1+\alpha^2)^{\frac{1}{2}}}{2Av^*}\right). \tag{4.31}$$

It is convenient to define a new dependent variable Y by

$$Y = \frac{\Omega r(1+\alpha^2)^{\frac{1}{2}}}{Av^*}; \tag{4.32}$$

then

$$\zeta_1 = e^{-\frac{1}{2}Y} \tag{4.33}$$

and, using equation (4.27),

$$\frac{\delta}{b} = Re_\phi^{-1}\frac{Ae^{-B/A}\,Y\,e^{Y}}{x(1+\alpha^2)^{\frac{1}{2}}}. \tag{4.34}$$

In terms of α and Y, the velocity components in the boundary layer may be expressed in the form

$$v_r = \begin{cases} \alpha\Omega r\left(1+\dfrac{1}{Y}\ln\zeta\right), & \zeta \le \zeta_1 \\ -\alpha\Omega r\dfrac{1}{Y}\ln\zeta, & \zeta_1 < \zeta \le 1 \end{cases} \tag{4.35}$$

$$v_\phi = -\Omega r\frac{1}{Y}\ln\zeta. \tag{4.36}$$

The integrals in equations (2.64) and (2.65) are

$$\int_0^\delta v_r^2\,dz = 2Ae^{-B/A}Re_\phi^{-1}\frac{\alpha^2F_1(Y)x}{(1+\alpha^2)^{\frac{1}{2}}}\Omega^2b^3, \tag{4.37}$$

$$\int_0^\delta v_rv_\phi\,dz = 2Ae^{-B/A}Re_\phi^{-1}\frac{\alpha F_2(Y)x}{(1+\alpha^2)^{\frac{1}{2}}}\Omega^2b^3, \tag{4.38}$$

$$\int_0^\delta v_\phi^2\,dz = 2Ae^{-B/A}Re_\phi^{-1}\frac{F_3(Y)x}{(1+\alpha^2)^{\frac{1}{2}}}\Omega^2b^3, \tag{4.39}$$

where

$$F_1(Y) = \frac{1}{Y}e^{\frac{Y}{2}}(e^{\frac{Y}{2}} - Y) \tag{4.40}$$

$$F_2(Y) = \frac{1}{Y} e^{\frac{Y}{2}} (e^{\frac{Y}{2}} - \tfrac{1}{2}Y - 2) \tag{4.41}$$

$$F_3(Y) = \frac{1}{Y} e^{Y}. \tag{4.42}$$

Using equations (4.02) and (4.32), the shear stress τ_0 is given by

$$\frac{\tau_0}{\rho} = \frac{\Omega^2 r^2 (1 + \alpha^2)}{A^2 Y^2}; \tag{4.43}$$

and, assuming (as in equation (4.12)) that $\tau_{r,0} = -\alpha\,\tau_{\varphi,0}$,

$$\frac{\tau_{r,0}}{\rho\Omega^2 b^2} = \frac{x^2 \alpha (1 + \alpha^2)^{\frac{1}{2}}}{A^2 Y^2}, \tag{4.44}$$

$$\frac{\tau_{\varphi,0}}{\rho\Omega^2 b^2} = -\frac{x^2 (1 + \alpha^2)^{\frac{1}{2}}}{A^2 Y^2}. \tag{4.45}$$

Equations (2.64) and (2.65) become

$$\frac{d}{dX}\left[X^2 \frac{\alpha^2}{(1 + \alpha^2)^{\frac{1}{2}}} F_1(Y)\right] - X \frac{\alpha}{(1 + \alpha^2)^{\frac{1}{2}}} F_3(Y) = -\frac{X^3 \alpha (1 + \alpha^2)^{\frac{1}{2}}}{2A^3 e^{-B/A} Y^2}, \tag{4.46}$$

$$\frac{d}{dX}\left[X^3 \frac{\alpha}{(1 + \alpha^2)^{\frac{1}{2}}} F_2(Y)\right] = -\frac{X^4 (1 + \alpha^2)^{\frac{1}{2}}}{2A^3 e^{-B/A} Y^2}, \tag{4.47}$$

where

$$X = x Re_{\varphi}^{\frac{1}{2}} \tag{4.48}$$

is a new independent variable (which is the square root of the local rotational Reynolds number).

Equations (4.46) and (4.47) are a pair of simultaneous equations for α and Y in terms of X. When their solution is found, equations (4.35) and (4.36) can be solved to determine the velocity components. The nondimensional mass flow-rate is then given by

$$\frac{\dot{m}_0}{\mu r} = 2\pi A e^{-B/A} \frac{\alpha}{(1 + \alpha^2)^{\frac{1}{2}}} e^{\frac{Y}{2}} (e^{\frac{Y}{2}} - 2), \tag{4.49}$$

and, using equation (4.32),

$$\frac{\tau_{\phi,0}}{\rho\Omega^2 b^2} = -\frac{(1+\alpha^2)^{\frac{1}{2}} x^2}{A^2 Y^2} . \tag{4.50}$$

(It should be noted that equation (2.74) is not valid because of the lack of similarity in the radial velocity profile.) For a boundary layer which is turbulent everywhere on the disc, the moment coefficient is

$$C_M = 8\pi A e^{-B/A} Re_\phi^{-1} \frac{\alpha_1}{(1+\alpha_1^2)^{\frac{1}{2}}} F_2(Y_1), \tag{4.51}$$

where α_1, Y_1 are respectively the values of α, Y when $x = 1$ $(r = b)$. The case in which there is transition from laminar to turbulent flow on the disc is discussed below.

Goldstein, in the original paper using this method, re-arranged equations (4.46) and (4.47) so that X and α were dependent variables depending on Y. He found an asymptotic solution for large Y in which

$$X^2 \sim 3A^3 e^{-B/A} Y e^{Y}, \qquad \alpha \sim \frac{1}{3} \tag{4.52}$$

and an implicit expression for C_M of the form

$$C_M^{-\frac{1}{2}} \sim a_1 \log_{10}(Re_\phi C_M^{\frac{1}{2}}) + a_2, \tag{4.53}$$

where

$$a_1 \simeq 1.414A, \quad a_2 \simeq 0.6141B - 2.828A \log_{10}(2.210A). \tag{4.54}$$

Goldstein did not use the pipe-flow values of A and B given in equation (4.03); instead he chose $a_1 = 1.97$ and $a_2 = 0.03$ to enable his results to fit the only data then available (those of Schmidt 1921 and of Kempf 1924). These values correspond to $A = 1.97$ and $B = 6.53$ which Goldstein assumed to be appropriate values for the rotating disc. It is now believed, however, that the experiments of Kempf are inaccurate for laminar flow, and it is therefore inadvisable to use them for turbulent flow; Goldstein's values of A and B are, therefore, likely to be in error.

Dorfman (1963) used an alternative method of solving equations (4.46) and (4.47). For three different ranges of the independent variable X defined in equation (4.48), he assumed that the functions $e^{-Y}F_1(Y)/Y$, $e^{-Y}F_2(Y)/Y$ and $e^{-Y}F_3(Y)/Y$ can be expressed as power-law distribibutions of Ye^{Y} within the range. He then solved the equations explicitly for each range and combined his solutions for $10^4 < Re_\phi < 10^9$ to give the equation

$$C_M = 0.491\left(\log_{10}(Re_\phi)\right)^{-2.58}. \tag{4.55}$$

With the modern computing techniques now available it is possible to compute the solution of equations (4.46) and (4.47) directly. They may be written in the form

$$\frac{2+\alpha^2}{\alpha(1+\alpha^2)}\frac{d\alpha}{dX} + \frac{F_1'(Y)}{F_1(Y)}\frac{dY}{dX} = -\frac{2}{X} + \frac{F_3(Y)}{\alpha X F_1(Y)} - \frac{X(1+\alpha^2)}{2A^3 e^{-B/A}\alpha Y^2 F_1(Y)}, \tag{4.56}$$

$$\frac{1}{\alpha(1+\alpha^2)}\frac{d\alpha}{dX} + \frac{F_1'(Y)}{F_2(Y)}\frac{dY}{dX} = -\frac{3}{X} - \frac{x(1+\alpha^2)}{2A^3 e^{-B/A}\alpha Y^2 F_2(Y)} \tag{4.57}$$

and then rearranged to give $d\alpha/dX$ and dY/dX explicitly in terms of X, α and Y. This is therefore an initial-value problem and starting values are required in order to integrate the equations. In addition, values of A and B must be specified: the values used for the computations were $A = 2.46$, $B = 5.4$ (the value of A is the same as that given in equation (4.03) for flow in a pipe). A variety of starting values has been used and the moment coefficient computed in each case: it was found found that, with the chosen values of A and B, all solutions were asymptotic to the curve given by equation (4.55).

In practice, the boundary layer on the disc is usually laminar for small values of x, and it is assumed that this laminar flow becomes unstable when $x = x_T$. As discussed in Section 4.1, transition to turbulent flow occurs at a critical value of the local rotational Reynolds number (depending on factors such as surface roughness); this is equivalent to a critical value, X_T, of the independent variable X.

To avoid infinite values of the stresses $\tau_{r,0}$ and $\tau_{\varphi,0}$ it is necessary that the integrals $\int_0^\infty v_r^2 dz$ and $\int_0^\infty v_r v_\varphi dz$ (which occur in the derivatives on the left-hand sides of equations (2.64) and (2.65)) are continuous when $x = x_T$. For laminar flow, these integrals can be computed using the results of Section 3.1.1 to give

$$\left[\int_0^\infty v_r^2 \, dz\right]_{lam} = 0.054098\Omega^2 b^3 Re_\varphi^{-\frac{1}{2}} x^2, \tag{4.58}$$

$$\left[\int_0^\infty v_r v_\varphi \, dz\right]_{lam} = 0.15398\Omega^2 b^3 Re_\varphi^{-\frac{1}{2}} x^2. \tag{4.59}$$

These expressions are equated to those for turbulent flow in equations (4.37) and (4.38) when $x = x_T$. This results in the simultaneous equations

$$\frac{\alpha_T^2 F_1(Y_T)}{(1+\alpha_T^2)^{\frac{1}{2}}} = \frac{0.054098}{2Ae^{-B/A}} X_T \tag{4.60}$$

$$\frac{\alpha_T F_2(Y_T)}{(1+\alpha_T^2)^{\frac{1}{2}}} = \frac{0.15398}{2Ae^{-B/A}} X_T \tag{4.61}$$

where α_T, Y_T are values of α, Y respectively at transition.

Appropriate values of X_T, α_T and Y_T for several transitional values of $x^2 Re_\varphi$ are shown in Table 4.2 and the resulting dependence of C_M on Re_φ is shown in

Table 4.2 Starting values for integration of the equations using a logarithmic profile: $A = 2.46$, $B = 5.4$

$x^2 Re_\varphi$	X_T	α_T	Y_T
5×10^4	224	7.35	0.370
7×10^4	265	7.53	0.369
10^5	316	7.73	0.368
2×10^5	447	8.11	0.366
3×10^5	548	8.33	0.365

Figure 4.2. The lowest value of $x^2 Re_\varphi$ at transition is that corresponding to the intersection of the laminar and the asymptotic turbulent curves: for values of the local rotational Reynolds number less than this, it is expected that no turbulent flow can occur. The highest value corresponds to the largest critical value obtained when a very smooth disc is used (see Section 4.1).

It should be noted that the curves which start from the laminar curve for C_M and finish on the asymptotic turbulent curve have been deduced assuming a logarithmic profile in the transition region. Although this is unlikely to be valid, it should give a reasonable estimate.

4.2.4 Numerical technique

Cebeci and Abbott (1975) found numerical solutions of equations (2.35) to (2.38), with $\partial p/\partial r = 0$; they used "effective viscosities" (see Section 2.21) with different formulations for the effective viscosity in the part of the boundary layer near the wall and that near the inviscid "core" outside the boundary layer. They also included a correction factor for the transitional region between laminar and turbulent flow. Using a grid of variable mesh, the equations of motion were converted to finite-difference equations which were then solved numerically.

Velocity profiles predicted by their method are shown in

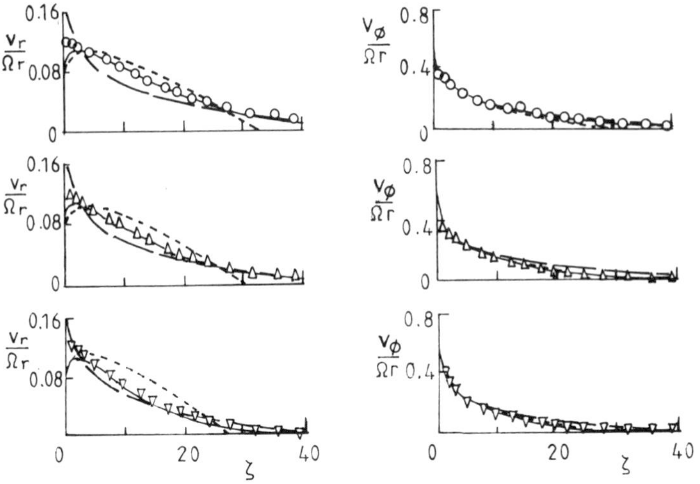

(a) $v_r/\Omega r$ (b) $v_\phi/\Omega r$

Figure 4.3 Turbulent velocity profiles

Theory:

———— numerical predictions (Cebeci and Abbott 1975)

— — — logarithmic profiles

- - - - - $\frac{1}{7}$-power-law profiles

Experiment: (Erian and Tong 1971)

▽ $Re_\phi = 5.6\text{x}10^5$, △ $Re_\phi = 7.6\text{x}10^5$, ○ $Re_\phi = 9.9\text{x}10^5$

Figure 4.3 for $Re_\phi = 5.6\text{x}10^5$, $7.6\text{x}10^5$ and $9.9\text{x}10^5$. Also shown in the figure are the $\frac{1}{7}$-power-law and logarithmic velocity profiles assumed for the theory presented in Sections 4.2.2 and 4.2.3.

Predictions of the moment coefficient have been carried out by Ong (1988) using the method of Cebeci and Abbott; her results are virtually indistinguishable from those using the logarithmic profile described in Section 4.2.3.

4.2.5 Comparison between theory and experiment

Using the apparatuses described in Section 3.1.2, Gregory, Stuart and Walker (1955) and Cham and Head (1969) made measurements of the velocity profiles for turbulent flow. The latter authors showed that the measured turbulent velocity profiles obeyed a logarithmic law near the disc, and they deduced the tangential component of the wall shear stress ($\tau_{\phi,0}$). Erian and Tong (1971), using an aluminium disc of 457 mm diameter rotating in air, employed hot-wire anemometry to measure the velocity profiles and the Reynolds stresses.

The numerical solutions of Cebeci and Abbott were in very good agreement with the velocity profiles measured by Cham and Head and by Erian and Tong, and the solutions were also in good agreement with the nondimensional shear stress measured by the former authors. Comparison between the numerical solutions and the measured velocity profiles of Erian and Tong are shown in Figure 4.3.

Schmidt (1921) and Kempf (1924) measured the moment coefficients for the free disc, and it was their results that Goldstein used to determine the constants in his logarithmic profile. As more recent work has suggested that these experimental results were inaccurate, they are not presented here.

Theodorsen and Regier (1944), in the experiments described in Section 3.1.2, measured the moment coefficients, using a number of different fluids, for rotational Reynolds numbers up to $Re_\phi = 7 \times 10^6$. Their measurements, which are shown in Figure 4.4, are in good agreement with equation (4.55). For their results, from an experiment in which transition occurred at approximately $x^2 Re_\phi = 3 \times 10^5$, there is also good agreement with the transitional curve found using a logarithmic profile as discussed in Section 4.2.3.

Owen (1969) obtained results for a disc of 762 mm diameter rotating in air at Reynolds numbers up to $Re_\phi \simeq 4 \times 10^6$. The moment coefficients for the free-disc case, measured by a torquemeter built into the drive

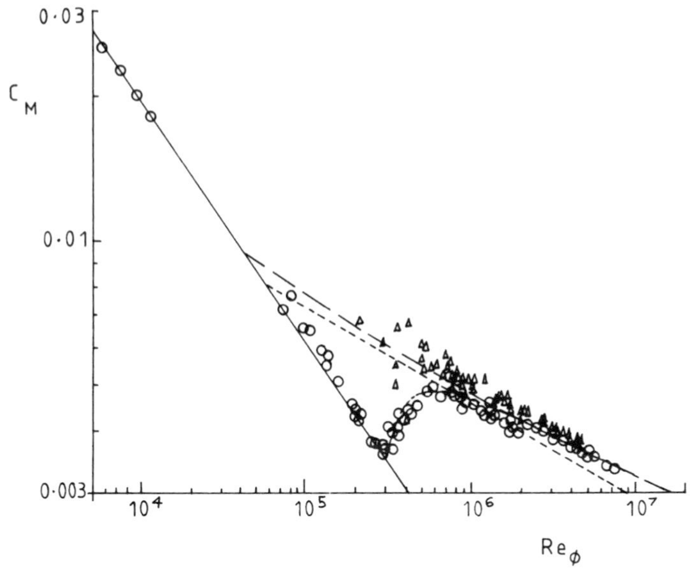

Figure 4.4 Variation of C_M with Re_ϕ

Theory:

———— laminar flow

— — — turbulent flow: asymptotic logarithmic profile

—·—· turbulent flow: transitional logarithmic profile

- - - - turbulent flow: $\frac{1}{7}$-power-law profile

Experiment:

○ Theodorsen and Regier (1944)

Δ Owen (1969)

---

shaft, are shown in Figure 4.4; it can be seen that they are consistently higher than the theoretical curve. (A slight change in the values for A and B in the theory of Section 4.2.3 would, of course, give an asymptotic curve in better agreement with Owen's data.) He correlated the moment coefficients by

$$C_M = 0.0655 Re_\phi^{-0.186}. \qquad (4.62)$$

Over the range $10^5 < Re_\phi < 10^7$, equations (4.55) and (4.62) differ by less than 5%. Even at Reynolds numbers as low as $Re_\varphi = 2.1 \times 10^5$, the moment coefficients showed no sign of transitional flow: atmospheric disturbances, produced by the air-cooled electric motor used to rotate the disc, were believed to have caused premature transition. Also, experimental uncertainty in the torque measurement was greater at the lower rotational speeds; this increased the scatter at the smaller values of Re_φ.

Although the $\frac{1}{7}$-power-law profile is often used to predict the moment coefficient, it provides a reasonable approximation to the turbulent-flow data shown in Figure 4.4 only for $10^5 < Re_\varphi < 10^6$, and it underestimates C_M at higher Reynolds numbers.

4.3 GENERALIZATION OF THE INTEGRAL EQUATIONS

4.3.1 A rotating disc in a rotating fluid

When $\frac{1}{7}$-power-law profiles are used for a disc rotating with angular speed Ω in a fluid rotating with speed $\omega = \beta\Omega$, equations (2.80) to (2.82) are valid with $v_{\varphi,0} = \Omega r$ and $v_{\varphi,\infty} = \beta\Omega r$. It is convenient to write

$$\frac{\delta}{r} = \gamma\,(x^2 Re_\varphi)^{-\frac{1}{5}}\,|1-\beta|^{\frac{3}{5}} \tag{4.63}$$

For solid-body rotation (β constant), equations (2.80) to (2.82) then become, after some manipulation,

$$\frac{\dot{m}_0}{\mu r} = \frac{49\pi}{60}\,\mathrm{sgn}(1-\beta)\,(x^2 Re_\varphi)^{\frac{4}{5}}|1-\beta|^{\frac{8}{5}}\alpha\gamma, \tag{4.64}$$

$$\frac{343}{1656}\left[\frac{18}{5} + 2\frac{x}{\alpha}\frac{d\alpha}{dx} + \frac{x}{\gamma}\frac{d\gamma}{dx}\right] - \frac{1+8\beta}{36(1-\beta)\alpha^2} = -0.0225\,\frac{(1+\alpha^2)^{\frac{3}{8}}}{(1-\beta)\alpha\gamma^{\frac{5}{4}}}, \tag{4.65}$$

$$\frac{49}{720}\left[\frac{23+37\beta}{5(1-\beta)} + \frac{x}{\alpha}\frac{d\alpha}{dx} + \frac{x}{\delta}\frac{d\gamma}{dx}\right] = 0.0225\,\frac{(1+\alpha^2)^{\frac{3}{8}}}{(1-\beta)\alpha\gamma^{\frac{5}{4}}}. \tag{4.66}$$

A solution of these equations exists with α and γ independent of x; α and γ, however, depend on β. This solution is given by

$$\alpha(\beta) = \left[\frac{2300(1 + 8\beta)}{49(1789 - 409\beta)}\right]^{\frac{1}{2}} \tag{4.67}$$

$$\gamma(\beta) = \left[\frac{81(1 + \alpha^2)^{\frac{3}{8}}}{49(23 + 37\beta)\alpha}\right]^{\frac{4}{5}}. \tag{4.68}$$

The solution clearly breaks down for $\beta \leq -\frac{1}{8}$ and for $\beta \geq 4.374$; variation of α and γ with β for $0 \leq \beta \leq 4$ is shown in Table 4.3.

Table 4.3 Variation of α, γ, ϵ_m, ϵ_m^r, ϵ_M and K_w with β

β	α	γ	ϵ_m	ϵ_m^r	ϵ_M	K_w
0.0	0.162	0.526	0.219	0.228	0.0729	0.383
0.1	0.220	0.368	0.175	0.171	0.0679	0.445
0.2	0.267	0.286	0.137	0.133	0.0604	0.507
0.3	0.309	0.233	0.105	0.103	0.0517	0.568
0.4	0.348	0.197	0.078	0.078	0.0425	0.630
0.5	0.385	0.170	0.055	0.057	0.0333	0.692
0.6	0.420	0.149	0.037	0.039	0.0243	0.753
0.7	0.454	0.132	0.022	0.025	0.0159	0.815
0.8	0.487	0.119	0.011	0.013	0.0086	0.877
0.9	0.520	0.108	0.004	0.003	0.0029	0.938
1.0	0.553	0.0983	0.000	-0.006	0.000	1.000
2.0	0.907	0.0498	-0.116	-0.111	0.163	1.617
3.0	1.445	0.0310	-0.349	-0.359	0.678	2.233
4.0	3.182	0.0200	-0.945	-0.723	2.341	2.850

The nondimensional mass flow-rate in the boundary layer is given by

$$\frac{\dot{m}_0}{\mu r} = \epsilon_m (x^2 Re_\phi)^{\frac{4}{5}} \tag{4.69}$$

where

$$\epsilon_m = \frac{49\pi}{60} \operatorname{sgn}(1 - \beta)\, \alpha\, \gamma\, |1 - \beta|^{\frac{8}{5}}, \tag{4.70}$$

and the moment coefficient is

$$C_M = \epsilon_M Re_\phi^{-\frac{1}{5}} \tag{4.71}$$

where

$$\epsilon_M = \frac{49\pi}{4140} \operatorname{sgn}(1 - \beta)\, |1 - \beta|^{\frac{8}{5}} (23 + 37\beta)\alpha\gamma. \tag{4.72}$$

Equation (2.74) showing the relationship between $\tau_{\phi,0}$, Re_ϕ and $\dot{m}_0$ is valid with the stress-mass-flow parameter given by

$$K_w = \frac{23 + 37\beta}{60}. \tag{4.73}$$

Values of ϵ_m, ϵ_M and K_w are shown in Table 4.3.

For use in the discussion of rotor-stator systems in Chapters 6 and 7, an approximation to ϵ_m is useful. This is given by $\epsilon_m \simeq \epsilon'_m$, where

$$\epsilon'_m = 0.154\beta^{\frac{8}{5}} - 0.386\beta^{\frac{4}{5}} + 0.228, \quad \beta < 1 \tag{4.74}$$

$$\epsilon'_m = -0.164\beta^{\frac{8}{5}} + 0.308\beta^{\frac{4}{5}} - 0.150, \quad \beta > 1. \tag{4.75}$$

Values of ϵ'_m are given in Table 4.3.

It is of interest to investigate the approximate solution using the Ekman integral equations (2.83) and (2.84). The axes are taken to be rotating with the angular speed Ω of the disc so that $v = 0$ at $z = 0$ and $v_c = -(1-\beta)\Omega r$. The relative velocity profiles are

$$u = -(1 - \beta)\alpha\Omega r\eta^{\frac{1}{7}}(1 - \eta), \qquad v = -(1 - \beta)\Omega r\eta^{\frac{1}{7}}, \tag{4.76}$$

and the shear stresses on the disc are given by equations (2.78) and (2.79) with $v_{\phi,0} - v_{\phi,c} = -(1-\beta)\Omega r$. It is elementary to show that

$$\alpha = 0.553, \quad \frac{\delta}{r} = 0.0983\,(1 - \beta)^{\frac{3}{5}}\,(x^2 Re_\phi)^{-\frac{1}{5}}, \tag{4.77}$$

$$\frac{\dot{m}_0}{\mu r} = 0.1395\,\mathrm{sgn}(1-\beta)\,|1 - \beta|^{\frac{8}{5}}\,(x^2 Re_\phi)^{\frac{4}{5}}. \tag{4.78}$$

This approximate solution is valid when $1-\beta$ is small, and the coefficients in equation (4.77) are identical with the values of α and γ shown in Table 4.3 for $\beta = 1$.

4.3.2 A stationary disc in a rotating fluid

For a fluid rotating with angular speed ω over a stationary disc, equations (2.78) to (2.82) are valid with $v_{\phi,0} = 0$ and $v_{\phi,\infty} = \omega r$. Writing

$$\frac{\delta}{r} = \gamma\,(x^2 Re'_\varphi)^{-\frac{1}{5}}, \tag{4.79}$$

where

$$Re'_\varphi = \frac{\rho\omega b^2}{\mu} \tag{4.80}$$

is a modified rotational Reynolds number. For ω constant, equations (2.80) to (2.82) then become, after some manipulation,

$$\frac{\dot{m}_0}{\mu r} = -\frac{49\pi}{60}\,(x^2 Re'_\varphi)^{\frac{4}{5}}\alpha\gamma \tag{4.81}$$

$$\frac{343}{1656}\left[\frac{18}{5} + 2\frac{x}{\alpha}\frac{d\alpha}{dx} + \frac{x}{\gamma}\frac{d\gamma}{dx}\right] + \frac{2}{9\alpha^2} = 0.0225\,\frac{(1+\alpha^2)^{\frac{3}{8}}}{\alpha\gamma^{\frac{5}{4}}} \tag{4.82}$$

$$\frac{49}{720}\left[-\frac{37}{5} + \frac{x}{\alpha}\frac{d\alpha}{dx} + \frac{x}{\gamma}\frac{d\gamma}{dx}\right] = -0.0225\,\frac{(1+\alpha^2)^{\frac{3}{8}}}{\alpha\gamma^{\frac{5}{4}}} \tag{4.83}$$

It is easy to verify that these equations have no solution with α constant (and real). It can be shown that, when $x \simeq 0$,

$$\alpha\gamma^5 \to \left(\frac{1444}{34300}\right)^4, \tag{4.84}$$

and

$$\alpha \to x^5\gamma, \quad \gamma = \frac{409}{3610}. \tag{4.85}$$

A solution may be obtained by a method similar to that used by Rott and Lewellen (1966). They assumed that, near a stationary disc, there is a boundary layer in which the fluid flows radially inward (just as in Section 3.2.2 for laminar flow); this boundary layer starts when $x = 1$ ($r = b$) with $\alpha = 0$ and $\delta = 0$. The equations have to be solved in the direction of flow, so a new variable

$$\xi = 1 - x \tag{4.86}$$

is defined, and it is assumed that the equations have a solution of the form

$$\alpha = \xi^{\frac{1}{2}} y_1(\xi), \qquad \gamma = \xi^{\frac{2}{5}} y_2(\xi), \tag{4.87}$$

where $y_1(\xi)$, $y_2(\xi)$ are functions such that $y_1(0) = \alpha_0$, $y_2(0) = \delta_0$; α_0 and δ_0 are nonzero constants which may be determined. (If the exponents of ξ in the definitions (4.87) are taken differently from the values given, the equations cannot be satisfied.)

In a rotor-stator system, fluid enters the boundary layer on the stator within a very small region near the periphery of the disc. For this case, therefore, it is not appropriate to assume that $\dot{m}_0$ is zero (which is implied in the theory of Rott and Lewellen) when $x = 1$. Nonzero values of α and δ need to be assumed and equations (4.82) and (4.83) can be integrated directly, starting from $x = 1$ and proceeding in a negative direction.

This procedure is not carried out here; instead the Ekman integral equations are solved exactly (as in the previous section), but with axes rotating with angular speed ω so that $v_0 = -\omega r$ and $v_\infty = 0$. This gives

$$\alpha = 0.553, \qquad \frac{\delta}{r} = 0.0983(x^2 Re_\phi')^{-\frac{1}{5}} \tag{4.88}$$

and

$$\frac{\dot{m}_0}{\mu r} = -0.1395(x^2 Re_\phi')^{\frac{4}{5}}. \tag{4.89}$$

which is of exactly the same form as before. There is, therefore, a solution of the approximate equations which is valid for solid-body rotation of the fluid (β constant). This solution is used with the rotor-stator systems discussed in Chapter 6.

Chew (1989) has proposed a modified form of the integral equations for a stationary disc. The functions $f(\zeta)$ and $g(\zeta)$ (see equations (4.07) and (4.08)) are not defined, but it is assumed that $\dot{m}_0$ is given by equation (4.13), as for a $\frac{1}{7}$-power-law velocity profile; the radial integral equation is not used but is replaced by the assumption that $\alpha = 0.364$; and the shape factor K_v (see equation (2.72)) is assumed to be $\frac{1}{12}$ instead of $\frac{1}{6}$ (as for a $\frac{1}{7}$-power-law velocity profile). (These assumptions are said to be consistent with numerical solutions for flow over the stator of a sealed rotor-stator system.)

4.3.3 Flow impinging on a rotating disc

Truckenbrodt (1954) investigated the boundary layer on a rotating disc with a flow normal to its surface where, as for the laminar case discussed in Section 3.2.3, $v_{r,\infty} = \sigma\Omega r$ and $v_{\phi,\infty} = 0$. Because there is a radial pressure gradient in this flow, the integral equation (2.60) contains extra terms and is replaced by

$$\frac{d}{dr}\left[r\int_0^\delta v_r(v_r - v_{r,\infty})dz\right] + r\frac{dv_{r,\infty}}{dr}\int_0^\delta (v_r - v_{r,\infty})dz - \int_0^\delta v_\phi^2 dz = -\frac{r}{\rho}\tau_{r,0}. \tag{4.90}$$

Equation (2.61) is unchanged (except that $v_{\phi,\infty} = 0$).

The velocity profiles are assumed to be of the form

$$v_r = \Omega r[\alpha(1 - \zeta) + \sigma\zeta]\zeta^{\frac{1}{n}}, \quad v_\phi = \Omega r(1 - \zeta^{\frac{1}{n}}), \tag{4.91}$$

which, when $n = 7$ and $\sigma = 0$, reduce to equations (4.07) and (4.08). Using the method described in Section 4.2.2, it can be shown that

$$\alpha^2 + R\alpha\sigma = S + T\sigma^2 \tag{4.92}$$

and

$$C_M = \gamma(G\alpha + H\sigma)Re_\phi^{-\frac{2}{n+3}}, \tag{4.93}$$

where G, H, R, S, T are tabulated in Table 4.4 and

$$\gamma = K^{\frac{n+1}{n+3}}\left[\frac{(n+3)(1+\alpha^2)^{\frac{n-1}{2(n+3)}}}{(5n+11)(G\alpha+H\sigma)}\right]^{\frac{n+1}{n+3}}, \tag{4.94}$$

where K is given in Table 4.1. Values of $C_M Re_\phi^{2/(n+3)}$ are tabulated in Table 4.5, and it can be shown that, as $\sigma \to \infty$,

$$C_M Re_\phi^{\frac{2}{n+3}} \sim 4\pi\gamma(G\alpha + H\sigma). \tag{4.95}$$

Table 4.4 Variation of G, H, R, S, T and Z with n

n	7	8	9	10
G	0.068	0.063	0.058	0.054
H	0.029	0.026	0.024	0.022
R	0.029	0.012	-0.002	-0.013
S	0.026	0.020	0.016	0.013
T	1.056	1.032	1.014	1.000
Z	0.113	0.085	0.067	0.054

Table 4.5 Variation of $C_M Re_\phi^{\frac{2}{n+3}}$ with n and σ

σ	$n=7$	$n=8$	$n=9$	$n=10$
0.0	0.073	0.056	0.045	0.037
0.1	0.079	0.061	0.049	0.040
0.2	0.086	0.066	0.053	0.043
0.3	0.093	0.071	0.057	0.046
0.4	0.099	0.076	0.061	0.049
0.5	0.106	0.081	0.065	0.052
0.6	0.112	0.086	0.068	0.055
0.7	0.119	0.091	0.072	0.058
0.8	0.125	0.096	0.076	0.062
0.9	0.132	0.101	0.080	0.065
1.0	0.139	0.106	0.084	0.068
2.0	0.210	0.161	0.129	0.104
3.0	0.280	0.217	0.173	0.141
4.0	0.348	0.270	0.217	0.177
5.0	0.414	0.322	0.260	0.212
6.0	0.477	0.373	0.301	0.247
7.0	0.538	0.422	0.341	0.280
8.0	0.598	0.470	0.381	0.313
9.0	0.657	0.517	0.420	0.346
10.0	0.714	0.563	0.458	0.378

4.4 EFFECT OF ROUGHNESS ON MOMENT COEFFICIENTS

For flow in pipes or over plates, roughness has no effect on the shear stress providing either the flow is laminar or the protrusions do not penetrate through the viscous sublayer (see Schlichting 1979). For turbulent flow, the thickness of the viscous sublayer, δ_* say, can be approximated by

$$\frac{\delta_* v^*}{\nu} \simeq 5, \tag{4.96}$$

where v^* is the friction velocity defined in equation (4.02). A surface is regarded as *hydraulically smooth* if $k_s < \delta_s$, where k_s is the size of the individual protrusions. (Most experiments were conducted with sand-roughened surfaces, and k_s was taken to be the size of the individual grains of sand used. Owing to the close proximity of the grains, the average height of the surface roughness was typically 60% of the size of the grains.)

It is convenient to classify turbulent flow into three regimes depending on the size of the roughness parameter, $k_s v^*/\nu$:

(i) *hydraulically-smooth* regime: $k_s v^*/\nu < 5$;

(ii) transition regime: $5 < k_s v^*/\nu < 70$;

(iii) *completely-rough* regime: $k_s v^*/\nu > 70$.

In regime (i), the shear stress depends only on the Reynolds number; in (iii) it depends only on the roughness; in (ii) it depends on both the Reynolds number and the roughness. For regime (iii), equation (4.01) can be replaced by

$$\frac{u}{v^*} = A \ln\left(\frac{z}{k_s}\right) + B \tag{4.97}$$

where

$$A = 2.5, \quad B = 8.5. \tag{4.98}$$

Dorfman (1958) applied this work to a completely rough free disc, and the theory is developed in the same way as that described in Section 4.2.3. For $2 \leq \ln(\delta/k_s) \leq 5$, Dorfman approximated $F_1(Y)$, $F_2(Y)$ and $F_3(Y)$ in equations (4.40) to (4.42) by

$$\frac{F_1(Y)}{Ye^Y} = 0.032\left(\frac{k_s}{\delta}\right)^{0.19}, \tag{4.99}$$

$$\frac{F_2(Y)}{Ye^Y} = 0.0335\left(\frac{k_s}{\delta}\right)^{0.19}, \tag{4.100}$$

$$\frac{F_3(Y)}{Ye^Y} = 0.057\left(\frac{k_s}{\delta}\right)^{0.28}, \tag{4.101}$$

where Y is now given by

$$Y = \frac{\Omega r(1+\alpha^2)^{\frac{1}{2}}}{Av^*} = \ln\left(\frac{\delta}{k_s}\right) + \frac{B}{A}. \qquad (4.102)$$

He then obtained solutions of the momentum-integral equations (4.46) and (4.47), and found, to a good approximation, that the moment coefficient is given by

$$C_M = 0.054\left(\frac{k_s}{b}\right)^{0.272}. \qquad (4.103)$$

Dorfman's equation (4.103) is in reasonable agreement with the measurements of Kanaev (1953) for a steel disc, with sand roughness $k_s/b = 3.33\text{x}10^{-3}$, rotating in water. However, the equation is not in agreement with Kanaev's data for a disc rotating in mercury, and it underestimates the measured moment coefficients of Theodorsen and Regier (1944), for $k_s/b = 8.3\text{x}10^{-4}$, by approximately 15%.

Granville (1973) also developed logarithmic relations for rough and smooth discs rotating in fluids that could contain drag-reducing polymer solutions. Whilst his theoretical result was able to predict accurately the measured effect of polymer addition on the moment coefficients, no such comparisons for the effect of surface roughness were presented.

CHAPTER 5
Heat Transfer from a Single Disc

The transfer of heat from a rotating disc which is at a different temperature from the surrounding fluid depends *inter alia* on the disc-temperature distribution, $T_0(x)$ and on Pr, the Prandtl number of the fluid. It is assumed below that $\partial T/\partial z \to 0$ as $z \to \infty$; for air, Pr is taken to be 0.71, unless stated to the contrary.

The special case (denoted by the subscript *spec*) in which $Pr = 1$ and the disc-temperature distribution is purely quadratic was discussed in Section 2.5.1. It was shown that the temperature, T_∞, outside the boundary layer depends on x unless the fluid is quiescent or there is a free vortex. It is assumed that the same is true for more general values of Pr and for other disc-temperature distributions.

The energy equation used to determine the temperature distribution within the boundary layer enables both the heat transfer from the disc and the temperature T_∞ to be determined. It is assumed throughout this chapter that there is a similarity solution for the temperature (or, when viscous dissipation is important, for the enthalpy); this is expected to be true for sufficiently large values of x (when the effect of the starting conditions have been attenuated).

For most of this chapter it is assumed that the disc-temperature distribution is of a power law form:

$$T_0(x) = T_{ref} + T_n x^n, \tag{5.01}$$

where T_{ref}, T_n and n are constants. When the Reynolds analogy is valid ($n = 2$ and $Pr = 1$), equation (2.106) shows that $H_\infty = H_{ref}$ (or, when dissipation is negligible, that $T_\infty = T_{ref}$) when $v_{\phi,\infty} = 0$. It is assumed throughout the rest of this chapter that this is also true when $Pr \neq 1$ or $n \neq 0$. More general distributions of $T_0(x)$ can be expressed as a power series in x; solutions based on equation (5.01) can be superposed for laminar flow, since the energy equation is then linear in T.

For laminar flow the temperature, $T(r,z)$, of the fluid satisfies the energy equation (2.23); for turbulent flow, and often for laminar flow, the approximate energy equation (2.39) is used instead of equation (2.23).

5.1 EXACT EQUATIONS FOR THE FREE DISC: LAMINAR FLOW

5.1.1 A solution of the energy equation

It was shown by Kibel' (1947) that, for laminar flow in a quiescent fluid with a disc-temperature distribution,

$$T_0(x) = T_1 + T_2 x^2, \tag{5.02}$$

where T_1 and T_2 are constants and $x = r/b$, there is a solution of the full equations of the form

$$T(x,\zeta) = T_\infty + \Theta_0(\zeta) + x^2 \Theta_2(\zeta), \tag{5.03}$$

where

$$\zeta = \left(\frac{\Omega}{\omega}\right)^{\frac{1}{2}} z \tag{5.04}$$

is the scaled coordinate used in Section 3.1.1 above. The functions $\Theta_0(\zeta)$ and $\Theta_2(\zeta)$ satisfy the equations

$$\Theta_0'' - Pr H \Theta_0' = -4\, Re_\phi^{-1}\, \Theta_2 - 12\, Pr\, Re_\phi \frac{\Omega^2 b^2}{c_p} F^2, \tag{5.05}$$

$$\Theta_2'' - Pr(H\Theta_2' + 2F\Theta_2) = -Re_\phi^{-1} \frac{\Omega^2 b^2}{c_p}(F'^2 + G'^2), \tag{5.06}$$

where primes denote differentiation with respect to ζ,

Re_ϕ is the rotational Reynolds number (see list of symbols) and F, G, H satisfy the von Kármán equations (3.05) to (3.07); the boundary conditions on Θ_0 and Θ_2 are

$$\Theta_0(0) = T_1 - T_\infty, \quad \Theta_0(\zeta) \to 0 \quad \text{as } \zeta \to \infty, \tag{5.07}$$

$$\Theta_2(0) = T_2, \quad \Theta_2(\zeta) \to 0 \quad \text{as } \zeta \to \infty. \tag{5.08}$$

It may be noted that equations (5.05) and (5.06) are linear, so that solutions may be superposed; further, equations (5.06) and (5.08) do not involve Θ_0 and so Θ_2 is independent of T_1. The solutions may be expressed in the form

$$\Theta_0(\zeta) = (T_1 - T_\infty)\alpha(\zeta) + \frac{\Omega^2 b^2}{c_p} Re_\phi^{-1} \beta(\zeta) + Re_\phi^{-1} T_2 \gamma(\zeta), \tag{5.09}$$

$$\Theta_2(\zeta) = T_2 \epsilon(\zeta) + \frac{\Omega^2 b^2}{c_p} \eta(\zeta), \tag{5.10}$$

where

$$\alpha'' - PrH\alpha' = 0, \tag{5.11}$$

$$\beta'' - PrH\beta' = -4\eta - 12PrF^2, \tag{5.12}$$

$$\gamma'' - PrH\gamma' = -4\epsilon, \tag{5.13}$$

$$\epsilon'' - Pr(H\epsilon' + 2F\epsilon) = 0, \tag{5.14}$$

$$\eta'' - Pr(H\eta' + 2F\eta) = -Pr(F'^2 + G'^2), \tag{5.15}$$

together with the boundary conditions

$$\alpha(0) = \epsilon(0) = 1, \quad \beta(0) = \gamma(0) = \eta(0) = 0, \tag{5.16}$$

$$\alpha(\zeta),\ \beta(\zeta),\ \gamma(\zeta),\ \epsilon(\zeta),\ \eta(\zeta) \to 0 \quad \text{as } \zeta \to \infty. \tag{5.17}$$

(It should be noted that the use of $\alpha(\zeta)$, $\beta(\zeta)$, $\gamma(\zeta)$, $\epsilon(\zeta)$ and $\eta(\zeta)$, as defined above, is exclusive to this section. In particular, β is used in other sections of this chapter to denote the core-swirl ratio $v_{\phi,\infty}/\Omega r$.)

For laminar flow, the heat flux from the disc is $q_0 = -k(\partial T/\partial z)_{z=0}$, so that

$$q_0 = -k\left(\frac{\Omega}{\nu}\right)^{\frac{1}{2}}\left[(T_1 - T_\infty)\alpha'(0) + Re_\varphi^{-1}\left(\frac{\Omega^2 b^2}{c_p}\beta'(0) + T_2\gamma'(0)\right) + x^2\left(T_2\epsilon'(0) + \frac{\Omega^2 b^2}{c_p}\eta'(0)\right)\right] \tag{5.18}$$

Determination of the derivatives $\alpha'(0)$, $\beta'(0)$, $\gamma'(0)$, $\epsilon'(0)$ and $\eta'(0)$ are, therefore, of particular importance.

When $Pr = 1$, it is easy to verify that equations (5.14) and (5.15) reduce to equations (3.05) and (3.06), with

$$\epsilon(\zeta) = G(\zeta),\ \eta(\zeta) = \tfrac{1}{2}G(\zeta) - \tfrac{1}{2}[F^2(\zeta) + G^2(\zeta)]; \tag{5.19}$$

further ϵ and η, as given in these equations, satisfy the boundary conditions given in equations (5.16) and (5.17). This result was obtained by Kibel' and is useful as a check on the numerical methods used for more general cases.

For values of Pr between 0.5 and 10, Millsaps and Pohlhausen (1952) developed a method for determining η for an isothermal disc (for which $T_2 = 0$). Sparrow and Gregg (1959) extended this work for values of Pr between 0.1 and 100 and neglected dissipative effects in their results. These authors used methods giving the solution in terms of quadratures, which are easy to perform when tabulated values of the functions $F(\zeta)$, $G(\zeta)$, $H(\zeta)$ and their derivatives are available. However, when Pr is very small, $\beta(\zeta)$ and $\gamma(\zeta)$ can be evaluated only by integration over a very large range of ζ and, unless great care is taken, large errors can occur. This appears to be the cause of the inaccurate values of $\beta(\zeta)$ computed by Millsaps and Pohlhausen; fortunately, for most cases of practical importance, this function has no significant effect on the heat transfer.

Since the numerical solution of differential equations with specified initial conditions is now much quicker than at the time when earlier workers produced their results, it is more satisfactory to solve the equations directly.

Table 5.1 Values of the derivatives $\alpha'(0)$, $\beta'(0)$, $\gamma'(0)$, $\epsilon'(0)$ and $\eta'(0)$ for different values of Pr.

Pr	$\alpha'(0)$	$\beta'(0)$	$\gamma'(0)$	$\epsilon'(0)$	$\eta'(0)$
0.1	-0.07658	1.10980	35.94254	-0.14173	0.04054
0.5	-0.26229	0.97096	6.30407	-0.42873	0.17126
0.71	-0.32586	1.01610	4.70465	-0.51848	0.23095
1.0	-0.39625	1.09790	3.67123	-0.61592	0.30796
2.0	-0.56526	1.38514	2.42391	-0.84662	0.54486
5.0	-0.85330	2.05090	1.55383	-1.23816	1.13185
10.0	-1.13412	2.82968	1.15567	-1.62057	1.93588
100.0	-2.68714	8.18291	0.48057	-3.74219	10.62056

(Note: β is not the core-swirl ratio)

This is particularly easy because equations (5.11) to (5.15) are linear and because $F'(0)$ and $G'(0)$ have already been determined to a high degree of accuracy (see equation (3.30)). The method requires special care for values of Pr less than unity, when the thermal boundary layer is thicker than the viscous one; an account of the method is given in Appendix B. Asymptotic values may be found for small and large values of Pr: these are of interest primarily as a check on the accuracy of computational methods and are discussed in Appendix C.

The values of $\alpha'(0)$, $\beta'(0)$, $\gamma'(0)$, $\epsilon'(0)$ and $\eta'(0)$ as functions of Pr are tabulated in Table 5.1, and the values of $\alpha(\zeta)$, $\beta(\zeta)$, $\gamma(\zeta)$, $\epsilon(\zeta)$ and $\eta(\zeta)$ for $Pr = 0.71$ in Table 5.2. (The equation for $\alpha(\zeta)$ was solved by Sparrow and Gregg, and their results are in agreement with those shown in Table 5.1.) Figure 5.1 shows the effect of Pr on the temperature profiles for *isothermal* ($T_2 = 0$) and *purely quadratic* ($T_1 = T_\infty$) disc-temperature distributions. It can be seen that, in both cases, the thickness of the thermal boundary layer increases as Pr decreases. In particular, for the purely quadratic case, the thermal boundary layer is thicker than the velocity one when $Pr < 1$ and thinner when $Pr > 1$ (since the temperature profile is identical with the tangential velocity profile when $Pr = 1$).

Table 5.2 Values of the functions $\alpha(\zeta)$, $\beta(\zeta)$, $\gamma(\zeta)$, $\epsilon(\zeta)$ and $\eta(\zeta)$: Pr = 0.71

ζ	$\alpha(\zeta)$	$\beta(\zeta)$	$\gamma(\zeta)$	$\epsilon(\zeta)$	$\eta(\zeta)$
0.0	1.0000	0.0000	0.0000	1.0000	0.0000
0.1	0.9674	0.1014	0.4508	0.9483	0.0209
0.2	0.9348	0.2018	0.8635	0.8972	0.0380
0.3	0.9023	0.3000	1.2398	0.8472	0.0517
0.4	0.8699	0.3947	1.5812	0.7987	0.0626
0.5	0.8376	0.4849	1.8893	0.7519	0.0711
0.6	0.8055	0.5697	2.1654	0.7071	0.0775
0.7	0.7737	0.6484	2.4109	0.6643	0.0822
0.8	0.7422	0.7204	2.6272	0.6237	0.0854
0.9	0.7112	0.7853	2.8158	0.5851	0.0874
1.0	0.6807	0.8429	2.9782	0.5487	0.0882
1.1	0.6507	0.8932	3.1157	0.5144	0.0883
1.2	0.6214	0.9364	3.2301	0.4820	0.0876
1.3	0.5928	0.9725	3.3227	0.4516	0.0863
1.4	0.5649	1.0019	3.3950	0.4231	0.0845
1.5	0.5378	1.0249	3.4486	0.3963	0.0824
1.6	0.5116	1.0419	3.4850	0.3712	0.0800
1.7	0.4861	1.0534	3.5056	0.3477	0.0773
1.8	0.4616	1.0597	3.5118	0.3256	0.0745
1.9	0.4380	1.0614	3.5051	0.3050	0.0716
2.0	0.4152	1.0589	3.4866	0.2857	0.0686
2.1	0.3934	1.0527	3.4576	0.2677	0.0657
2.2	0.3725	1.0431	3.4194	0.2508	0.0627
2.3	0.3524	1.0305	3.3730	0.2350	0.0597
2.4	0.3333	1.0154	3.3194	0.2202	0.0568
2.5	0.3150	0.9980	3.2597	0.2064	0.0540
2.6	0.2976	0.9788	3.1948	0.1935	0.0512
2.7	0.2810	0.9579	3.1253	0.1814	0.0485
2.8	0.2652	0.9358	3.0522	0.1701	0.0459
2.9	0.2502	0.9125	2.9761	0.1595	0.0435
3.0	0.2360	0.8885	2.8977	0.1495	0.0411
3.1	0.2225	0.8638	2.8174	0.1402	0.0388
3.2	0.2097	0.8386	2.7359	0.1315	0.0366
3.3	0.1976	0.8131	2.6536	0.1234	0.0346
3.4	0.1862	0.7875	2.5710	0.1157	0.0326
3.5	0.1753	0.7619	2.4882	0.1086	0.0307
3.6	0.1651	0.7363	2.4058	0.1019	0.0289
3.7	0.1554	0.7109	2.3240	0.0956	0.0272
3.8	0.1463	0.6858	2.2431	0.0897	0.0257
3.9	0.1376	0.6610	2.1631	0.0842	0.0241
4.0	0.1295	0.6366	2.0845	0.0790	0.0227
4.2	0.1146	0.5893	1.9315	0.0696	0.0201
4.4	0.1013	0.5440	1.7851	0.0613	0.0178
4.6	0.0896	0.5010	1.6460	0.0540	0.0157
4.8	0.0791	0.4605	1.5145	0.0476	0.0139
5.0	0.0699	0.4225	1.3909	0.0420	0.0123
∞	0	0	0	0	0

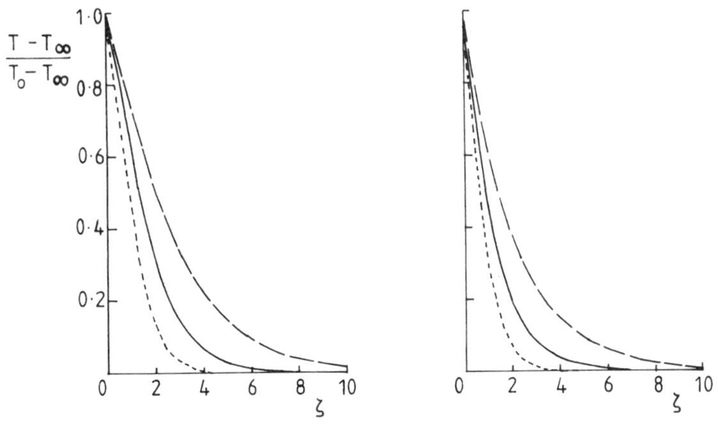

(a) isothermal (b) purely quadratic

Figure 5.1 Temperature profiles for the free disc: laminar flow

— — — $Pr = 0.5$, ———— $Pr = 1$, - - - - $Pr = 2$

5.1.2 Nusselt numbers

The local Nusselt number is defined as

$$Nu = \frac{rq_0}{k(T_0 - T_\infty)}, \tag{5.20}$$

where q_0 is given by equation (5.18). This is consistent with the definition given in the list of symbols, taking

$$q_w = q_0, \quad \Delta T = T_0(x) - T_\infty = T_1 - T_\infty + T_2 x^2, \tag{5.21}$$

T_1, T_2 being the constants occurring in equation (5.02). Hence

$$\frac{Nu}{(x^2 Re_\varphi)^{\frac{1}{2}}} = -\frac{T_1 - T_\infty}{\Delta T}[\alpha'(0) - \epsilon'(0)] - \epsilon'(0)$$
$$- \frac{T_2}{\Delta T} Re_\varphi^{-1} \gamma'(0) - \frac{\Omega^2 b^2}{c_p \Delta T}[Re_\varphi^{-1} \beta'(0) + x^2 \eta'(0)], \tag{5.22}$$

where ΔT is given by equation (5.21). The ratio $\Omega^2 b^2 / c_p \Delta T$

which occurs in equation (5.22) is related to Ec, the Eckert number (see list of symbols). It is easy to verify that the average Nusselt number Nu_{av} (see list of symbols) is given by

$$\frac{Nu_{av}}{Re_\phi^{\frac{1}{2}}} = -\frac{T_1 - T_\infty}{T_1 - T_\infty + \frac{1}{2}T_2}[\alpha'(0) - \epsilon'(0)] - \epsilon'(0)$$
$$- \frac{T_2}{T_1 - T_\infty + \frac{1}{2}T_2} Re_\phi^{-1}\gamma'(0) - \frac{T_1 - T_\infty + T_2}{T_1 - T_\infty + \frac{1}{2}T_2}[Re_\phi^{-1}\beta'(0) + \tfrac{1}{2}\eta'(0)]Ec_b, \qquad (5.23)$$

where Ec_b is the value of Ec when $x = 1$ ($r = b$).

In many cases of interest, $Ec << 1$ (that is, dissipative effects are small), and the terms involving β and η are negligible. In such cases, two special disc-temperature distributions are of interest. For an isothermal disc $T_2 = 0$; hence equations (5.22) and (5.23) give

$$Nu_{iso} = -\alpha'(0)(x^2 Re_\phi)^{\frac{1}{2}}, \quad Nu_{av,iso} = -\alpha'(0) Re_\phi^{\frac{1}{2}}, \qquad (5.24)$$

where $\alpha'(0)$ is given in Table 5.1. The variation of $Nu_{iso}/(x^2 Re_\phi)^{\frac{1}{2}}$ with Pr is shown in Figure 5.2.

For a purely quadratic disc-temperature distribution (so that $T_1 = T_\infty$), it follows that, for $Re_\phi >> 1$,

$$Nu_{quad} = -\epsilon'(0)(x^2 Re_\phi)^{\frac{1}{2}}, \quad Nu_{av,quad} = -\epsilon'(0) Re_\phi^{\frac{1}{2}}, \qquad (5.25)$$

where $\epsilon'(0)$ is given in Table 5.1. Figure 5.2 shows the variation of $Nu_{quad}/(x^2 Re_\phi)^{\frac{1}{2}}$ with Pr.

For the case of an adiabatic disc ($q_0 = 0$), $T_0 = T_{0,ad}$ where $T_{0,ad}$ is the adiabatic-disc temperature (see Section 2.5.3), which is given by

$$T_{0,ad} = T_{1,ad} + T_{2,ad} x^2. \qquad (5.26)$$

From equation (5.18) with $q_0 = 0$, it follows that

$$T_{1,ad} = T_\infty + \frac{\gamma'(0)\eta'(0) - \epsilon'(0)\beta'(0)}{\alpha'(0)\epsilon'(0)} Re_\phi^{-1} \frac{\Omega^2 b^2}{c_p}, \qquad (5.27)$$

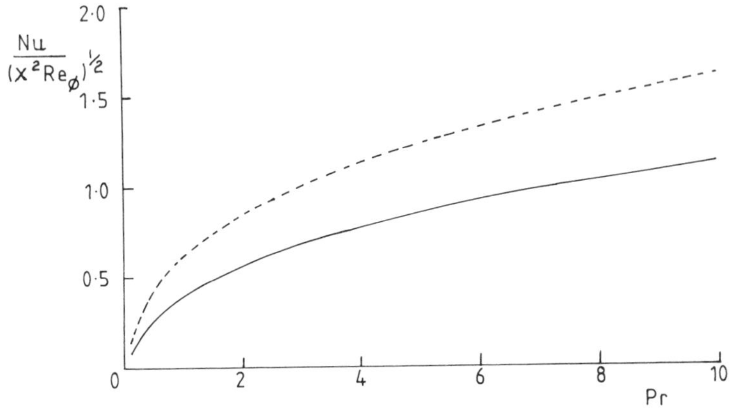

Figure 5.2 Variation of $Nu/(x^2 Re_\phi)^{\frac{1}{2}}$ with Pr for the free disc: laminar flow

——— isothermal, - - - - purely quadratic

$$T_{2,ad} = -\frac{\eta'(0)}{\epsilon'(0)}\frac{\Omega^2 b^2}{c_p}. \qquad (5.28)$$

If $Re_\phi >> 1$ then $T_{1,ad} \simeq T_\infty$ and the quadratic term in the expression for $T_{0,ad}$ dominates (at least for large values of x). Equation (5.26) is then approximated by

$$T_{0,ad} = T_\infty + R\frac{\Omega^2 r^2}{2c_p}, \qquad (5.29)$$

where R, the recovery factor (see Section 2.5.5), is

$$R = -2\frac{\eta'(0)}{\epsilon'(0)} \qquad (5.30)$$

Values of R obtained from this equation are given in Table 5.3, together with comparisons with the quantities $Pr^{1/3}$ and $Pr^{1/2}$ (the latter value is quoted by Schlichting 1979 as being appropriate for laminar flow over a stationary flat plate with Prandtl number of order unity). It is clear that, at least in the neighbourhood of $Pr = 1$, a good approximation to R for laminar free-disc flow is given by $Pr^{1/3}$.

Table 5.3 Values of the recovery factor, R, for the free disc: laminar flow

Pr	0.1	0.5	0.71	1.0	2.0	5.0	10.0	100.0
R	0.572	0.799	0.891	1.000	1.287	1.828	2.389	5.676
$Pr^{1/3}$	0.464	0.794	0.892	1.000	1.260	1.710	2.154	4.642
$Pr^{1/2}$	0.316	0.707	0.843	1.000	1.414	2.236	3.162	10.00

At high rotational speeds, *frictional heating* can cause the surface temperature of a thermally-insulated disc to be much higher than that of the surrounding fluid, and a nonadiabatic disc will only be cooled if $T_0 > T_{0,ad}$. If $T_{0,ad} > T_0 > T_\infty$, the disc will be heated by the "cooler" fluid, and the Nusselt number based on the definition given in equation (5.20) becomes negative! This negative value may be avoided by using the Nusselt number, Nu^*, defined in equation (2.116); this is based on $\Delta T = T_0 - T_{0,ad}$ instead of on $\Delta T = T_0 - T_\infty$.

When the disc-temperature distribution is purely quadratic ($T_1 = T_\infty$) and $Re_\phi >> 1$, equations (5.18) and (5.28) give

$$q_0 \simeq \left(\frac{\Omega}{\nu}\right)^{\frac{1}{2}} \epsilon'(0)(T_2 - T_{2,ad})x^2, \tag{5.31}$$

and $T_0 - T_{0,ad} \simeq (T_2 - T_{2,ad})x^2$. It follows that

$$Nu^*_{quad} = -\epsilon'(0)\,(x^2 Re_\phi)^{\frac{1}{2}}, \quad Nu^*_{av,quad} = -\epsilon'(0)\,Re_\phi^{\frac{1}{2}}. \tag{5.32}$$

These expressions are the same as those for Nu_{quad} and $Nu_{av,quad}$ given in equation (5.25) for the case where dissipation is negligible.

5.2 SOLUTION OF THE BOUNDARY-LAYER EQUATIONS: LAMINAR FLOW

It is possible to extend the theory of Section 5.1 to more general disc-temperature distributions by expanding the solution in powers of r^2; this has been done by Tifford (1951). However, it is usually preferable to use the boundary-layer approximations, and it is assumed throughout the rest of this chapter that these approximations have been made; the temperature profile is then determined from equation (2.39) instead of from equation (2.23).

Since the temperature outside the boundary layer cannot be prescribed, except when the fluid there is quiescent (or there is a free vortex), any theory should predict both the Nusselt number Nu and the temperature distribution T_∞. The discussion here is restricted to the case in which the fluid outside the boundary layer is in solid-body rotation (or at rest), so that $\beta = v_\phi/\Omega r$ is constant.

(The meaning of the symbol β has now reverted to the core-swirl ratio, as throughout most of the book. The exceptional use, in Section 5.1, of the function $\beta(\zeta)$ is no longer relevant.)

5.2.1 Differential energy equation

It is convenient to consider first disc-temperature distributions of the form given by equation (5.01). The temperature distribution within the boundary layer is assumed to be

$$T(x,\zeta) = T_{ref} + T_n x^n \theta(\zeta), \tag{5.33}$$

where $\zeta = (\Omega/\nu)^{\frac{1}{2}} z$, as before, and $\theta(\zeta)$ is a function of ζ such that $\theta(0) = 1$ and $\partial\theta/\partial\zeta \to 0$ as $\zeta \to \infty$. The local Nusselt number Nu (based on $\Delta T = T_0 - T_{ref}$) and the temperature distribution, T_∞, outside the boundary layer are then given by

$$Nu = -\theta'(0)\,(x^2 Re_\phi)^{\frac{1}{2}}, \qquad T_\infty(x) = T_{ref} + T_n \theta_\infty x^n, \tag{5.34}$$

where θ_∞ is the limiting value of $\theta(\zeta)$ as $\zeta \to \infty$.

Substitution of the relation (5.33) into equation (2.39) gives

$$\theta'' - Pr\,(H\theta' + nF\theta) = -Pr\,Ec\,\theta_\infty(F'^2 + G'^2); \qquad (5.35)$$

in this equation $F(\zeta)$, $G(\zeta)$ and $H(\zeta)$ are the non-dimensional components of velocity for the flow discussed in Chapter 3, and Ec is the Eckert number based on the temperature difference $\Delta T = T_0 - T_{ref} = T_n x^n$. The term on the right-hand side of equation (5.35) represents the effect of viscous dissipation on the heat transfer: the inclusion of this term is valid only if Ec is independent of x, which is true only when $n = 2$.

The boundary conditions on $\theta(\zeta)$ are given by

$$\theta(0) = 1, \quad \theta'(\zeta) \to 0 \text{ as } \zeta \to \infty. \qquad (5.36)$$

For the free disc (for which it is assumed that T_∞ is constant and $T_{ref} = T_\infty$), it is possible to apply the more stringent conditions

$$\theta(0) = 1, \quad \theta(\zeta) \to 0 \text{ as } \zeta \to \infty. \qquad (5.37)$$

If the conditions (5.37) are valid, so are those of equation (5.36); the advantage of equations (5.37) is that accurate numerical solutions are easier to obtain, especially for small values of Pr. This is discussed in more detail in Appendix B.

When $n = 2$, it follows that $\Delta T \propto x^2$ and Ec is constant; it is easy to verify that the solution of equation (5.35) subject to the boundary conditions (5.36) is

$$\theta(\zeta) = \epsilon(\zeta) + Ec\,\eta(\zeta), \qquad (5.38)$$

where $\epsilon(\zeta)$ and $\eta(\zeta)$ satisfy equations (5.14) and (5.15) repectively, together with the boundary conditions

$$\epsilon(0) = 1,\ \eta(0) = 0,\ \epsilon'(\zeta) \to 0,\ \eta'(\zeta) \to 0 \quad \text{as } \zeta \to \infty \qquad (5.39)$$

The relation (5.34) for the Nusselt number then gives

$$Nu = -[\epsilon'(0) + Ec\,\eta'(0)]\,(x^2 Re_\phi)^{\frac{1}{2}}. \qquad (5.40)$$

The expression for Nu^* is the same as that given in equation (5.32).

For the free disc, when the conditions (5.37) can be used, $\epsilon(\zeta)$ amd $\eta(\zeta)$ satisfy the relevant boundary conditions of equations (5.16) and (5.17); in this case, the values of $\epsilon'(0)$ and $\eta'(0)$ are given in Table 5.1.

The special case for which $n = 2$ and $Pr = 1$ can be simplified further, using equation (5.19). Equation (5.38) then becomes

$$\theta_{spec}(\zeta) = G(\zeta) + \tfrac{1}{2} Ec\,[G(\zeta) - \{F^2(\zeta) + G^2(\zeta)\}]. \qquad (5.41)$$

It follows (since $F(0) = 0$ and $G(0) = 1$) that

$$\theta_{\infty, spec} = \beta[1 + \tfrac{1}{2}Ec(1 - \beta)], \qquad (5.42)$$

$$Nu_{spec} = - G'(0)(1 - \tfrac{1}{2}Ec)(x^2 Re_{\phi})^{\frac{1}{2}}, \qquad (5.43)$$

where $G'(0)$ is given in Table 3.2 for several values of β. It is easy to show that the results agree with those produced by the Reynolds analogy discussed in Section 2.5.

If the disc-temperature distribution is not quadratic (that is, when $n \neq 2$), the dissipation term cannot be included in equation (5.35). Neglecting dissipation (which, in laminar flow, is usually small), the equation reduces to

$$\theta'' - Pr\,(H\theta' + nF\theta) = 0, \qquad (5.44)$$

which may be solved in a manner similar to that described for $\epsilon(\zeta)$ in Appendix B. When $n = 2$, the solution is identical with the function $\epsilon(\zeta)$ specified by equation (5.10); when $n = 0$, the solution is given by $\alpha(\zeta)$ specified by equation (5.09); when $n = -2$, equation (5.44) can be solved directly (using equation (3.05)) and the heat transfer is identically zero.

It is easy to demonstrate that, using the definitions given in equation (5.20),

$$\frac{Nu}{(x^2 Re_\phi)^{\frac{1}{2}}} = \frac{Nu_{av}}{Re_\phi^{\frac{1}{2}}} = -\theta'(0). \qquad (5.45)$$

Values of these quantities are given in Table 5.4 for several values of Pr and of n, and Figure 5.3 shows their variation with n for $Pr = 0.71$.

The negative values of Nu for $n < -2$ in Table 5.4 can be understood by considering the case of a cold fluid flowing over a hot disc. The fluid close to the disc surface

Table 5.4 Values of $Nu/(x^2 Re_\phi)^{\frac{1}{2}} = Nu_{av}/Re_\phi^{\frac{1}{2}}$ for the free disc: laminar flow

Pr	$n=-3$	$n=-2$	$n=-1$	$n=0$	$n=1$	$n=2$	$n=3$
0.1	-0.0438	0	0.0399	0.0766	0.1104	0.1417	0.1709
0.5	-0.2113	0	0.1590	0.2623	0.3531	0.4287	0.4935
0.71	-0.2952	0	0.1893	0.3259	0.4319	0.5185	0.5918
1.0	-0.4074	0	0.2352	0.3962	0.5180	0.6159	0.6982
2.0	-0.7692	0	0.3482	0.5653	0.7226	0.8466	0.9498
5.0	-1.715	0	0.5445	0.8533	1.070	1.238	1.377
10.0	-3.057	0	0.7368	1.134	1.408	1.621	1.796

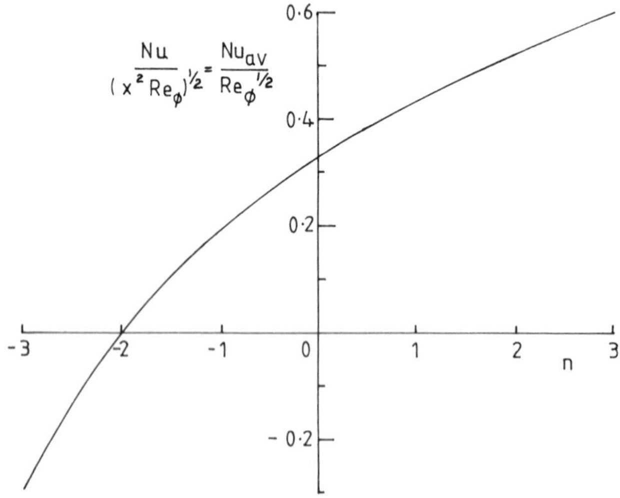

Figure 5.3 Variation of $Nu/(x^2 Re_\phi)^{\frac{1}{2}} = Nu_{av}/Re_\phi^{\frac{1}{2}}$ with n for the free disc: laminar flow, $Pr = 0.71$

decreases in temperature, as it moves radially outward in the boundary layer, at a slower rate than the disc itself. As a consequence, heat is transferred to the disc despite the fact that the disc is at a higher temperature than the fluid outside the boundary layer!

It is convenient to write

$$Nu = Pr\chi\, Nu_{Pr=1}, \tag{5.46}$$

where $Nu_{Pr=1}$ is the value of Nu for the given disc-temperature distribution when $Pr = 1$. (The correction factor χ is the same as that in Section 2.5.5.) The values of $Pr^{\frac{1}{2}}\chi$ for several values of n and Pr are shown in Table 5.5. For values of Pr near to unity (including $Pr = 0.71$) and $n \simeq 2$, it is convenient to write $\chi = \chi_{quad}$, where

$$\chi_{quad} \simeq Pr^{-\frac{1}{2}}. \tag{5.47}$$

Table 5.5 Values of $Pr^{\frac{1}{2}}\chi$ for the free disc: laminar flow

Pr	$n=-3$	$n=-2$	$n=-1$	$n=0$	$n=1$	$n=2$	$n=3$
0.1	0.3400	1	0.5365	0.6114	0.6740	0.7275	0.7740
0.5	0.7335	1	0.9560	0.9363	0.9640	0.9844	0.9996
0.71	0.8599	1	0.9552	0.9762	0.9895	0.9991	1.0059
1.0	1	1	1	1	1	1	1
2.0	1.3351	1	1.0468	1.0089	0.9864	0.9720	0.9619
5.0	1.8826	1	1.0353	0.9632	0.9238	0.8989	0.8820
10.0	2.3729	1	0.9906	0.9051	0.8596	0.8323	0.8134

For temperature distributions more general than those described in equation (5.34), since the equation for the temperature is linear, solutions may be superposed. Hence the disc-temperature distribution can be expanded in powers of x; each term of the expansion is then solved separately, using the above method, and the solutions are added.

The variation of $Nu/Re_\phi^{\frac{1}{2}}$ with x is shown for two temperature distributions in Figure 5.4: both distributions are cubics designed to be approximately equivalent

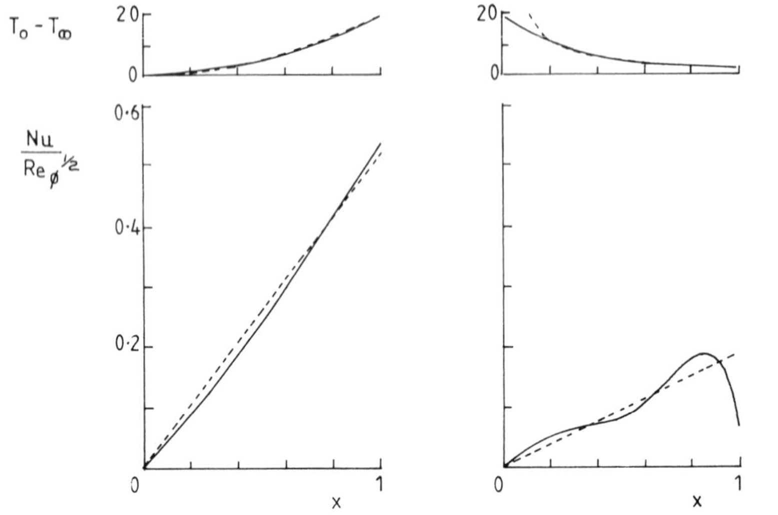

(a) (b)

Figure 5.4 Variation of $Nu/Re_\phi^{\frac{1}{2}}$ with x for the free disc: laminar flow, $Pr = 0.71$

	Curve	$T_0 - T_\infty$	$Nu_{av}/Re_\phi^{\frac{1}{2}}$
(a)	————	cubic	0.5191
	- - - - -	power law ($n = 2$)	0.5185
(b)	————	cubic	0.2040
	- - - - -	power law ($n = -1$)	0.1893

to power-law distributions. $T_0 - T_\infty$ is approximately proportional to x^2 in Figure 5.4a, and in Figure 5.4b to x^{-1} (except for small values of x). It can be seen that, in each case, the dependence of $Nu/Re_\phi^{\frac{1}{2}}$ on x for the cubic and for the approximating power law are not identical. The appropriate values of $Nu_{av}/Re_\phi^{\frac{1}{2}}$ are given in the figure; there is, in each case, reasonable agreement between the value for the cubic and that for the approximating power law.

When $\beta \neq 0$, equation (5.41) is unchanged, but the functions F and H must be computed using the equations discussed in Section 3.2.1. Work is in progress to compute $\theta'(0)$ and θ_∞ for a range of values of Pr, n and β; this will be reported by Rogers (1989).

5.2.2 Energy-integral equation for the free-disc case

An alternative approach is to use an integral technique similar to that described by Dorfman (1963) for a disc with an arbitrary temperature distribution rotating in a quiescent fluid at constant temperature T_∞. When viscous dissipation is negligible (as it usually is for laminar flow), it is assumed that the Nusselt number defined using $\Delta T = T_0 - T_\infty$ is given by

$$Nu = F_T(Pr)\, x^2 Re_\phi \left[\int_0^\delta \frac{2\pi v_r (T - T_\infty)}{\nu (T_0 - T_\infty)}\, dz \right]^{-\sigma'} \qquad (5.48)$$

where F_T is a function of Pr and σ' is a constant. Neither $F_T(Pr)$ nor σ' depends on x or on the form of the temperature profile in the boundary layer; each is different, however, for laminar and turbulent flow. (The function $F_T(Pr)$ is eliminated during the course of the following argument and it is not necessary to determine it explicitly.)

It is convenient to define a *temperature shape factor* K_T, by analogy with the "velocity shape factor" K_v defined in equation (2.72), as

$$K_T = \frac{\int_0^\delta v_r (T - T_\infty)\, dz}{(T_0 - T_\infty) \int_0^\delta v_r\, dz}. \qquad (5.49)$$

This ratio depends on the form of the temperature profile in the boundary layer and, therefore, on the Prandtl number. (When the conditions for the Reynolds analogy hold, $K_T = K_v$.) Equation (5.49) then becomes

$$Nu = F_T(Pr)\, x^2 Re_\phi \left(K_T \frac{\dot{m}_0}{\mu r}\right)^{-\sigma'} . \tag{5.50}$$

Equation (2.127), with $T_1 = T_\infty$, $Ec = 0$ and $\beta = 0$, gives

$$Nu = \frac{1}{\pi} K_w Pr \chi \frac{\dot{m}_0}{\mu r}, \tag{5.51}$$

where K_w is the stress-flow-rate parameter given by equation (2.74). Comparison of equation (5.50) with (5.51) gives

$$\frac{1}{\pi} K_w Pr \chi \left(\frac{\dot{m}_0}{\mu r}\right)^{\sigma' + 1} = F_T(Pr)\, x^2 Re_\phi K_T^{-\sigma'} . \tag{5.52}$$

For similarity solutions of the momentum equations, it is often assumed that

$$\frac{\dot{m}_0}{\mu r} = \epsilon_m (x^2 Re_\phi)^\sigma \tag{5.53}$$

where the "similarity parameters" ϵ_m and σ are independent of x (see, for example, equations (3.28) and (4.21)). Equations (2.74), (5.50) and (5.51) then give (when $\beta = 0$)

$$K_w = (\sigma + \tfrac{3}{2}) K_v , \qquad \sigma' = \frac{1 - \sigma}{\sigma}. \tag{5.54}$$

For laminar flow, using equation (3.42), it follows that

$$\sigma_{lam} = \frac{1}{2}, \quad \epsilon_{m,\,lam} = -\pi H_\infty, \quad K_{w,\,lam} H_\infty = G'(0), \quad \sigma'_{lam} = 1, \tag{5.55}$$

where $G'(0) = -0.6159$ and $H_\infty = -0.8845$ (see equation (3.23)). (The argument is continued below with general values of K_w, ϵ_m and σ so that other applications of the method can more easily be effected.)

Since dissipation is negligible, and T_∞ is constant, equation (2.87) can be written in the form

$$\frac{d}{dx}\left[x \int_0^\delta v_r (T - T_\infty)\, dz\right] = \nu \frac{Nu}{Pr} (T_0 - T_\infty). \tag{5.56}$$

Using the definition (5.49) and equations (5.51) and (5.52), it follows that

$$\frac{d}{dx}\left[K_T(T_0 - T_\infty)x^{2\sigma+1}\right] = 2K_w \chi (T_0 - T_\infty)x^{2\sigma}, \qquad (5.57)$$

since ϵ_m is independent of x.

For the special case in which $Pr = 1$ and $T_0 - T_\infty \propto x^2$, this is an identity since $\chi = 1$, $K_T = K_v$ and $2K_w = (2\sigma + 3)K_v$; this is to be expected, since the Reynolds analogy is valid. More generally, a relationship between K_T and χ is required in order to solve equation (5.54); a method for determining such a relationship is described below.

First the case in which $T_0 - T_\infty \propto x^2$ but $Pr \neq 1$ is considered. In this case it is convenient to write χ and K_T as χ_{quad} and $K_{T,quad}$ respectively; clearly both χ_{quad} and $K_{T,quad}$ depend on Pr, and it is assumed that neither depends on x. It follows from equation (5.57) that

$$K_{T,quad} = K_v \chi_{quad} \qquad (5.58)$$

in general and, using equation (5.47),

$$\chi_{quad,lam} = Pr^{-\frac{1}{2}}. \qquad (5.59)$$

Further, equation (5.52) gives

$$\frac{1}{\pi} K_w Pr \chi_{quad} \left(\frac{\dot{m}_0}{\mu r}\right)^{\sigma' + 1} = F_T(Pr)\, x^2 Re_\phi K_{T,quad}^{-\sigma'}. \qquad (5.60)$$

For the case with an arbitrary disc-temperature distribution, equation (5.52) holds and this, together with equation (5.60), gives

$$\frac{\chi}{\chi_{quad}} = \left(\frac{K_T}{K_{T,quad}}\right)^{-\sigma'}, \qquad (5.61)$$

where σ' is given by equation (5.55). Equations (5.58) and (5.61) can be combined to give

$$K_T = K_v \chi_{quad} \left(\frac{\chi}{\chi_{quad}}\right)^{-\frac{\sigma}{1-\sigma}}. \qquad (5.62)$$

Substitution of this expression into equation (5.57) leads to an equation for χ of the form

$$\frac{d}{dx}\left[\left(\frac{\chi}{\chi_{quad}}\right)^{-\frac{\sigma}{1-\sigma}}(T_0 - T_\infty)\, x^{2\sigma+1}\right] = \qquad (5.63)$$

$$(2\sigma + 3)\frac{\chi}{\chi_{quad}}(T_0 - T_\infty)\, x^{2\sigma}.$$

After some rearrangement, this equation can be integrated directly to give

$$\frac{\chi}{\chi_{quad}} = \left(\frac{\sigma}{2\sigma+3}\right)^{1-\sigma}\left[x^{2\sigma+1}(T_0 - T_\infty)\right]^{\frac{1-\sigma}{\sigma}}\left[\Psi_0(x)\right]^{-(1-\sigma)}, \qquad (5.64)$$

where

$$\Psi_0(x) = \int_0^x (T_0 - T_\infty)^{\frac{1}{\sigma}} x^{\frac{1+\sigma}{\sigma}} dx. \qquad (5.65)$$

Using equations (5.51) and (5.53), it follows that

$$Nu = \qquad (5.66)$$

$$\frac{K_v \epsilon_m \sigma}{2\pi}\left(\frac{2\sigma+3}{\sigma}\right)^{\sigma} Pr\,\chi_{quad}(x^2 Re_\varphi)^{\sigma}\left[x^{2\sigma+1}(T_0 - T_\infty)\right]^{\frac{1-\sigma}{\sigma}}\left[\Psi_0(x)\right]^{-(1-\sigma)}.$$

The average Nusselt number, Nu_{av}, can be obtained by using equation (2.134) with $T_1 = T_\infty$. This gives

$$Nu_{av} = \frac{K_v \epsilon_m}{\pi} Pr\,\chi_{quad}\left(\frac{2\sigma + 3}{\sigma}\right)^{\sigma} Re_\varphi^{\sigma}\,[\Psi_0(1)]^{\sigma}\,(T_0 - T_\infty)_{av}^{-1}. \qquad (5.67)$$

For laminar flow, using equations (5.64) and (5.67),

$$\chi_{lam} = \frac{1}{2\sqrt{2}} Pr^{-\frac{1}{2}} x^2 (T_0 - T_\infty)\left[\int_0^x (T_0 - T_\infty)^2 x^3 dx\right]^{-\frac{1}{2}}, \qquad (5.68)$$

$$Nu = -\frac{G'(0)}{2\sqrt{2}} Pr^{\frac{1}{2}} (x^2 Re_\varphi)^{\frac{1}{2}} x^2 (T_0 - T_\infty)\left[\int_0^x (T_0 - T_\infty)^2 x^3 dx\right]^{-\frac{1}{2}}; \qquad (5.69)$$

where $G'(0) = 0.6159$ (see equation (3.23)). In particular, when $T_0 - T_\infty \propto x^n$, this gives

$$Nu = \frac{1}{2}(n+2)^{\frac{1}{2}} G'(0) Pr^{\frac{1}{2}} (x^2 Re_{\varphi})^{\frac{1}{2}}, \quad Nu_{av} = Nu_b. \tag{5.70}$$

Table 5.6 shows a comparison of the results given by this equation with those of the solution of equation (5.45). (It is apparent from equation (5.70) that the method breaks down when $n < -2$; physically this is due to the fact that when n is sufficiently negative, the temperature profile in the boundary layer may not be monotonic. If this happens, K_T is not necessarily positive and equation (5.50) is meaningless: this implies that the basic assumption, equation (5.48), is invalid.)

Table 5.6 Ratio of Nu computed using equation (5.69) to Nu computed using equation (5.45)

Pr	$n=-1$	$n=0$	$n=1$	$n=2$	$n=3$
0.1	0.41	0.56	0.66	0.73	0.79
0.5	0.73	0.85	0.94	0.99	1.01
0.71	0.73	0.89	0.96	1.00	1.02
1.0	0.76	0.91	0.97	1	1.01
2.0	0.80	0.92	0.96	0.97	0.98
5.0	0.79	0.88	0.90	0.90	0.89
10.0	0.76	0.82	0.84	0.83	0.83

5.2.3 Integral method for a rotating disc in a rotating fluid

When $\beta \neq 0$, the temperature outside the boundary layer is not, in general, known *a priori* (see Section 2.5.1). It is, therefore, convenient to use $\Delta T = T_0 - T_{ref}$ in the definition of the Nusselt number so that, using equation (2.128) with $Ec_{ref} = 0$,

$$Nu = \frac{1}{\pi} K_w Pr \, \chi' \, \frac{\dot{m}_0}{\mu r}. \tag{5.71}$$

The integral method described in Section 5.2.2 can be extended to the case in which β is constant; this will be described by Rogers (1989).

5.2.4 Numerical solution

Ong (1988) extended the numerical solution of the differential boundary-layer equations discussed in Section 4.2.4 by including the energy equation. For laminar flow, in which viscous dissipation is negligible, her computed values of the local Nusselt numbers for $Pr = 0.1$, 1, 10 and $-3 \leq n < 3$ are in good agreement with the results presented in Table 5.4: most values agree to 3 significant figures. Taking the Prandtl number of air to be 0.72, her computations gave $Nu_{av}/Re_{\phi}^{\frac{1}{2}} = 0.329$ for an isothermal disc, which is less than 1% higher than the value given in Table 5.4 for $Pr = 0.71$.

5.3 SOLUTION OF THE BOUNDARY-LAYER EQUATIONS FOR TURBULENT FLOW

5.3.1 The Reynolds analogy

The Reynolds analogy (see Section 2.5) is valid if $Pr = 1$, the disc-temperature distribution is quadratic and the total enthalpy outside the boundary layer, H_{∞}, is independent of x. In this case equations (2.121) and (2.122) can be combined with equations (4.21) to (4.23) respectively to give (for a $\frac{1}{7}$-power-law velocity profile)

$$Nu^* = 0.0267\,(x^2 Re_{\phi})^{\frac{4}{5}}, \quad Nu^*_{av} = 0.0232\,Re_{\phi}^{\frac{4}{5}}. \qquad (5.72)$$

5.3.2 Integral method (neglecting dissipation)

The method described for laminar flow in Sections 5.2.2 and 5.2.3 can be applied to turbulent flow; the values of ϵ_m and K_w are given by equations (4.70) and (4.74) respectively and K_v is easily computed using equations (2.76) and (2.77). It follows that, when $\beta = 0$,

$$\sigma = \frac{4}{5}, \quad K_v = \frac{1}{6}, \quad K_w = \frac{23}{60}, \quad \epsilon_m = 0.2186. \qquad (5.73)$$

The dependence of χ_{quad} on Pr is essentially an empirical one. Schlichting (1979) gives several expressions for the corresponding factor for turbulent flow over a flat plate, and the arguments producing these expressions can be extended to the case of a rotating disc. In this case X_{quad} depends on both Pr and Re_ϕ, but it is usually found that it is adequate, at least for Pr of order unity, to take

$$X_{quad,\,turb} = Pr^{-\frac{2}{5}}. \tag{5.74}$$

Equations (5.66) and (5.67) then give

$$Nu = \tag{5.75}$$
$$0.0188 Pr^{\frac{3}{5}} (x^2 Re_\phi)^{\frac{4}{5}} x^{\frac{13}{20}} (T_0 - T_\infty)^{\frac{1}{4}} \left[\int_0^x (T_0 - T_\infty)^{\frac{5}{4}} x^{\frac{9}{4}} dx \right]^{-\frac{1}{5}},$$

$$Nu_{av} = 0.0470 Pr^{\frac{3}{5}} Re_\phi^{\frac{4}{5}} (T_0 - T_\infty)_{av}^{-1} \left[\int_0^1 (T_0 - T_\infty)^{\frac{5}{4}} x^{\frac{9}{4}} dx \right]^{\frac{4}{5}}. \tag{5.76}$$

For a power-law disc-temperature distribution such as that given in equation (5.01), $T_{ref} = T_\infty$ (since $\beta = 0$) and equations (5.75) and (5.76) simplify to

$$Nu = 0.0267 \left(\frac{n + 2.6}{4.6} \right)^{\frac{1}{5}} Pr^{\frac{3}{5}} (x^2 Re_\phi)^{\frac{4}{5}}, \quad Nu_{av} = \frac{n + 2}{n + 2.6} Nu_b, \tag{5.77}$$

where Nu_b is the value of Nu where $x = 1$ $(r = b)$. From the expression for Nu in equation (5.77), it can be seen that the method breaks down for $n \lesssim 2.6$. Table 5.7 gives the values of $Nu/(x^2 Re_\phi)^{\frac{4}{5}}$ and $Nu_{av}/Re_\phi^{\frac{4}{5}}$ for a selection of values of Pr and n.

Computations were also carried out for the two cubic disc-temperature distributions shown in Figure 5.4 for the laminar case. For turbulent flow, the values of Nu obtained were indistinguishable, when presented graphically, from the those found for the approximating power-law distribution. As for the laminar flow case, the method can be extended for $\beta \neq 0$; this is discussed by Rogers (1989).

Table 5.7 Values of $Nu/(x^2 Re_\phi)^{\frac{4}{5}}$ and $Nu_{av}/Re_\phi^{\frac{4}{5}}$ for the free disc: turbulent flow, dissipation neglected

(a) $Nu/(x^2 Re_\phi)^{\frac{4}{5}}$

Pr	$n=-2$	$n=-1$	$n=0$	$n=1$	$n=2$	$n=3$
0.1	0.0045	0.0054	0.0060	0.0064	0.0067	0.0070
0.5	0.0117	0.0142	0.0157	0.0168	0.0176	0.0183
0.71	0.0145	0.0176	0.0194	0.0207	0.0217	0.0226
1.0	0.0177	0.0216	0.0238	0.0254	0.0267	0.0277
2.0	0.0269	0.0327	0.0361	0.0385	0.0404	0.0421
5.0	0.0466	0.0567	0.0625	0.0667	0.0701	0.0729
10.0	0.0707	0.0860	0.0947	0.1011	0.1062	0.1104

(b) $Nu_{av}/Re_\phi^{\frac{4}{5}}$

Pr	$n=-2$	$n=-1$	$n=0$	$n=1$	$n=2$	$n=3$
0.1	0.0000	0.0034	0.0046	0.0053	0.0058	0.0062
0.5	0.0000	0.0089	0.0121	0.0140	0.0153	0.0163
0.71	0.0000	0.0110	0.0149	0.0172	0.0189	0.0202
1.0	0.0000	0.0135	0.0183	0.0212	0.0232	0.0248
2.0	0.0000	0.0205	0.0277	0.0321	0.0352	0.0375
5.0	0.0000	0.0354	0.0481	0.0556	0.0609	0.0651
10.0	0.0000	0.0537	0.0729	0.0843	0.0923	0.0986

5.3.3 Integral method (with dissipation)

For flow in a rotating cavity, Chew and Rogers (1988) extended the method of Section 5.3.2 to include the effect of dissipation. The *enthalpy shape factor* K_H is defined as

$$K_H = \frac{\int_0^\delta v_r (H - H_\infty)\, dz}{(H_0 - H_\infty)\int_0^\delta v_r\, dz} \tag{5.78}$$

by analogy with the "temperature shape factor" K_T (see equation (5.49)). Their basic assumption is that equations (5.58) and (5.61) are valid with K_T replaced by K_H; then equation (5.62) is replaced by

$$K_H = K_v\, \chi_{quad} \left(\frac{\chi}{\chi_{quad}}\right)^{-\frac{\sigma}{1-\sigma}} . \tag{5.79}$$

When $\beta = 0$ the energy equation (2.86), together with equations (2.74) and (2.122), can be written in the form

$$\frac{d}{dx}\left(x\int_0^{\delta} v_r (H - H_\infty)dz\right) = \nu \frac{K_w}{\pi}\frac{\dot{m}_0}{\mu r}\left[\chi(H_0 - H_\infty) + [1 - \tfrac{1}{2}\chi(1+R)]\Omega^2 r^2\right]. \tag{5.80}$$

It follows that

$$\frac{d}{dx}\left[K_H (H_0 - H_\infty)x^{2\sigma+1}\right] = 2K_w x^{2\sigma}\left[\chi(H_0 - H_\infty) + [1 - \tfrac{1}{2}\chi(1+R)]\Omega^2 r^2]\right]. \tag{5.81}$$

The recovery factor, R, is usually taken as

$$R = Pr^{\frac{1}{3}}, \tag{5.82}$$

the empirical value appropriate to the discussion of compressible flow (see Schlichting 1979). Equation (5.81) can then be solved as an initial value problem for the ratio χ/χ_{quad} (see Appendix D).

The dependence of $Nu^*/Re_\phi^{\frac{4}{5}}$ on x is shown in Figure 5.5 for disc-temperature distributions of the form given in equation (5.01) for $Pr = 0.71$, for $n = 2$, 0, -1, and for $Ec_b = 0$, 2, where Ec_b is the value of Ec evaluated at $x = 1$ ($r = b$). The difference between the values of Nu^* for $Ec_b = 0$ and $Ec_b = 2$ at $x = 1$ (where the difference is greatest) is less than 2% when $n = 2$, and less than 12% when $n = -2$; it should be remembered, however, that the accuracy of the method decreases as $|n - 2|$ increases. (It is of interest to note that the effect of dissipation when $Ec_b = 2$ is less than twice that when $Ec_b = 1$.)

The relationship between Nu^* (based on the temperature difference $\Delta T = T_0 - T_{0,ad}$) and Nu (based on $\Delta T = T_0 - T_{ref}$) is given by equation (2.131). When $Ec_b = 0$, $Nu^* = Nu$; when $Ec_b = 2$ and $Pr = 0.71$,

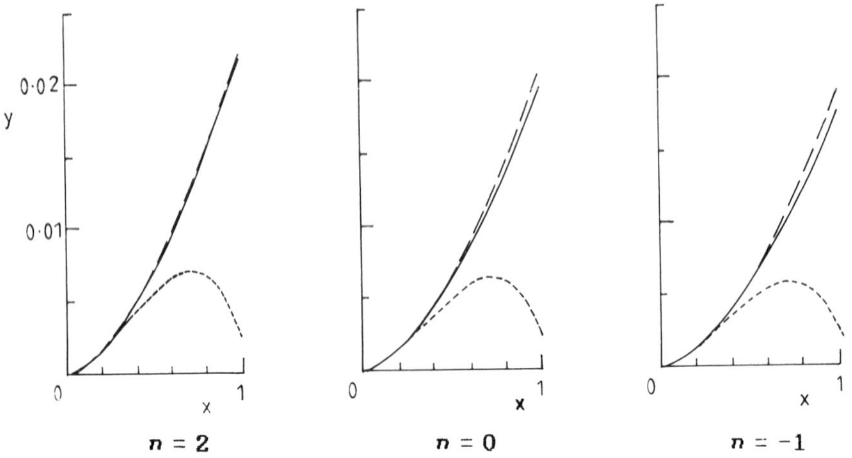

Figure 5.5 Effect of Eckert number on the variation of Nusselt number with x for the free disc: turbulent flow, $Pr = 0.71$

———— $y = Nu/Re_\varphi^{\frac{4}{5}} = Nu^*/Re_\varphi^{\frac{4}{5}}$: $Ec_b = 0$

— — — — $y = Nu^*/Re_\varphi^{\frac{4}{5}}$: $Ec_b = 2$

- - - - - - - $y = Nu/Re_\varphi^{\frac{4}{5}}$: $Ec_b = 2$

$$Nu = (1 - 0.89x^2)Nu^* \tag{5.83}$$

where, since $\beta = 0$, it is assumed that $R' = R$ (see Section 2.5.5) and that R is given by equation (5.82). The variation of $Nu/Re_\varphi^{\frac{4}{5}}$ with x is shown in Figure 5.5 and comparison between the curves for $Ec_b = 0$ and $Ec_b = 2$ shows that the effect of Ec_b on Nu is very much greater than the effect on Nu^*. Thus, to a good approximation, values of Nu^* obtained, experimentally or theoretically, at one value of Ec_b can be used at other values: the same cannot be said for Nu. It follows that it is better to use Nu^*, rather than Nu, for the Nusselt number, as recommended in Section 2.5.5. (When $Ec_b \ll 1$, so that dissipation is negligible, $Nu^* \simeq Nu$ and the problem does not arise.)

5.3.4 Numerical solution

As stated in Section 5.2.4, Ong (1988) obtained numerical solutions of the differential boundary-layer momentum and energy equations. For turbulent flow she used the Cebeci and Smith (1974) eddy-viscosity concept (see Section 4.2.4), incorporating a transitional factor and a turbulent Prandtl number, Pr_{turb} (see Section 2.5) of 0.9.

The variation of Nu^*_{av} with Re_ϕ predicted by Ong for the free disc, in the isothermal and the quadratic case, is shown in Figure 5.6. Also shown are the curves obtained using equation (5.77) (with Nu_{av} replaced by Nu^*_{av}) for $Pr = 0.72$ and $n = 0,\ 2$. At the larger values of Re_ϕ, where the flow is expected to be turbulent over most of the surface of the disc, the agreement between the numerical and integral results is very good.

Comparison with experimental data is discussed below.

5.4 EXPERIMENT

(Unless otherwise stated, the experiments described below were conducted in air with the disc rotating about a horizontal axis.)

Cobb and Saunders (1956) conducted experiments with a disc of 457 mm diameter with a built-in electric heater and a guard heater at the periphery to maintain a uniform temperature. Average Nusselt numbers were determined over the range $10^4 < Re_\phi < 6\text{x}10^5$, and transition from laminar to turbulent flow occurred at $Re_\phi \simeq 2.4\text{x}10^5$. For $Re_\phi < 2\text{x}10^5$, the results were correlated by

$$Nu_{av} = 0.36\, Re_\phi^{\frac{1}{2}}, \tag{5.84}$$

which is about 10% higher than the expression given in equation (5.24) for laminar flow with $Pr = 0.71$.

It should be noted that, in laminar flow, the effects of both radiation and natural convection may be important. When the ratio of the centripetal acceleration of the disc

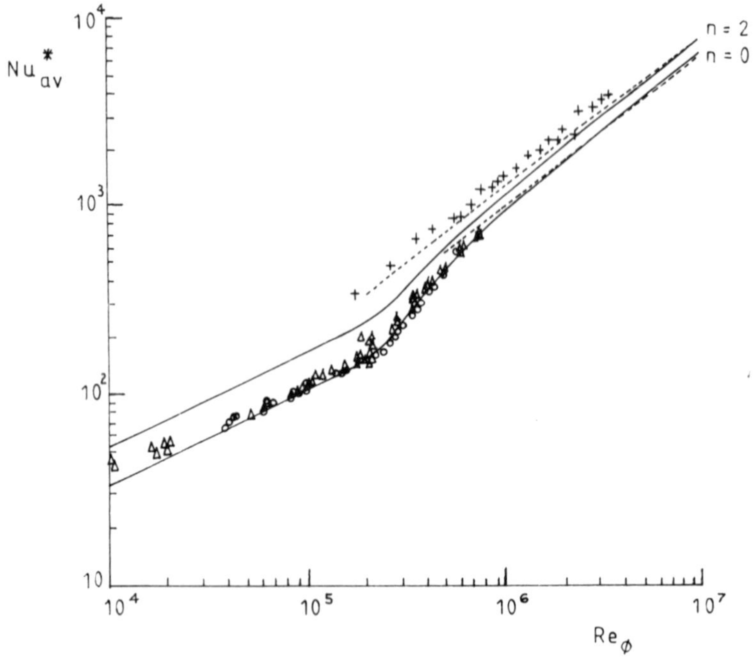

Figure 5.6 Variation of Nu^*_{av} with Re_ϕ for the free disc: $Pr = 0.72$

———— numerical solutions (Ong 1988)

- - - - - integral solutions, equation (5.77)

Experiment:

isothermal disc: $(n = 0)$
△ Cobb and Saunders (1956)
∘ McComas and Hartnett (1970)

nonisothermal disc: $(n \simeq 2)$
+ Owen, Haynes and Bayley (1974)

$(\Omega^2 r)$ to the gravitational acceleration of the earth is sufficiently small, natural convection can increase the Nusselt numbers significantly. Correlations to account for this effect have been proposed by Richardson and Saunders (1963), Mabuchi, Tanaka and Sakakibara (1971) and Popiel and Boguslawski (1975).

Kreith, Taylor and Chong (1959) used a mass-transfer

analogy. They rotated a disc of 203 mm diameter, coated with naphthalene, about a vertical axis in air for $3.5\times10^4 < Re_\varphi < 7\times10^5$. By weighing the disc before and after rotation, the authors deduced the average Sherwood number, Sh_{av}: this is analogous to an average Nusselt number for an isothermal disc rotating in a fluid with a Prandtl number of 2.4. Using equations (5.45) to (5.47), for $n = 0$ and $Pr = 2.4$,

$$Sh_{av} = Nu_{av} = 0.614 Re_\varphi^{\frac{1}{2}}. \tag{5.85}$$

This provides a good fit, for $Re_\varphi < 2\times10^5$, to the authors' data.

Nikitenko (1963), using a horizontal isothermal disc and apparatus described in Section 8.1, made measurements of local Nusselt numbers. His results for laminar flow were about 7% higher than those given in Table 5.4; for turbulent flow, his results were in good agreement with those of Table 5.7.

McComas and Hartnett (1970) made heat-transfer measurements on a disc of 483 mm diameter. Seven built-in electric heaters were used to maintain a uniform temperature, and the average Nusselt numbers were measured for $2\times10^4 < Re_\varphi < 6\times10^5$ from a heat balance or, for laminar flow, from the measured temperature gradient of the air near the disc. Figure 5.7 shows the measured distributions of $(T - T_\infty)/(T_0 - T_\infty)$ as a function of $\zeta = z(\Omega/\nu)^{\frac{1}{2}}$ for laminar flow; the theoretical curve (which corresponds to $\alpha(\zeta)$ in Section 5.1.1) is in good agreement with the experimental data.

The variation of Nu_{av} with Re_φ (using the results of Cobb and Saunders and of McComas and Hartnett) is shown in Figure 5.6. It should be pointed out that these authors used Nu_{av} rather than Nu^*_{av}; for their experiments, the difference is negligible. An extrapolated correlation for turbulent flow, given by

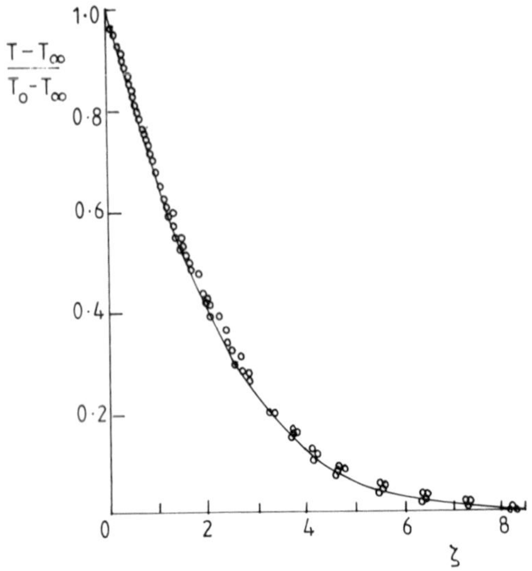

Figure 5.7 Temperature profiles for an isothermal free disc: laminar flow, $Pr = 0.71$

——— Theory, Section 5.1.1

∘ Experiment (McComas and Hartnett 1970)

$$Nu_{av} = 0.015 Re_\phi^{\frac{4}{5}}, \tag{5.86}$$

was proposed by Cobb and Saunders; this correlation is in close agreement with the results shown in Table 5.7b (for $Pr = 0.71$ and $n = 0$). The numerical results of Ong (see Section 5.3.4) are in good agreement with the experimental data.

Owen, Haynes and Bayley (1974) measured the average Nusselt numbers for $2\text{x}10^5 < Re_\phi < 4\text{x}10^6$ for a disc of 762 mm diameter. The disc was heated by stationary radiant heaters, and this resulted in a nonuniform temperature distribution that the authors approximated by a quadratic form ($n \simeq 2$). Using the measured surface temperatures as boundary conditions, they solved Laplace's conduction equation to determine the temperature distribution inside the disc; this was then used to calculate the surface heat flux. They correlated their results by

$$Nu^*_{av} = 0.0171 Re_\phi^{0.814}, \tag{5.87}$$

which agrees with the Reynolds analogy applied to Owen's (1969) measured moment coefficients, which are shown in Figure 4.4. Their measured Nusselt numbers, like the moment coefficients, are higher than the theoretical curves (see Figure 5.6). As discussed in Section 4.2.5, there was no sign of transition even at Reynolds numbers as low as $Re_\phi = 2.1 \times 10^5$.

Northrop and Owen (1988) made measurements on an internally-heated disc of 950 mm diameter rotating at Reynolds numbers up to $Re_\phi = 3.2 \times 10^6$. With five independently-controlled electric heaters, it was possible to generate a range of temperature distributions; these were represented by power-law profiles with $n \simeq -0.2$, 0.1, 0.4 and 0.6. Local heat-fluxes were determined from the solution of Laplace's conduction equation and from fluxmeters embedded in the disc surface. The results from the fluxmeters are shown in Figure 5.8, for $n \simeq -0.2$ and 0.6, together with Ong's numerical solution. The agreement between the computed and measured values of Nu^* is good except at the lower values of Re_ϕ, where radiation and natural convection were believed to have a significant effect on the experimental measurements. Ong noted that viscous dissipation (which she included in the energy equation) caused a large reduction in heat transfer at high values of Re_ϕ.

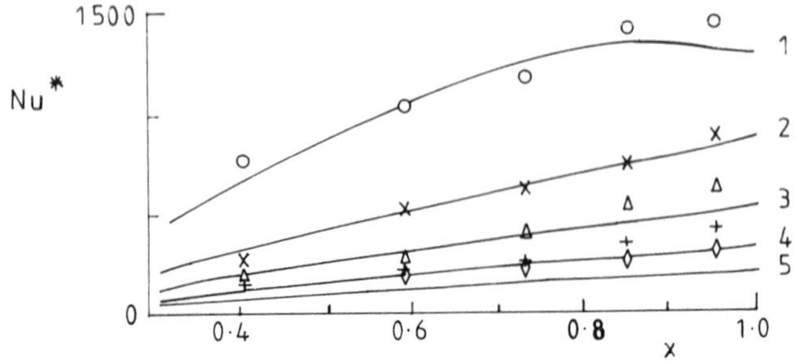

(a)

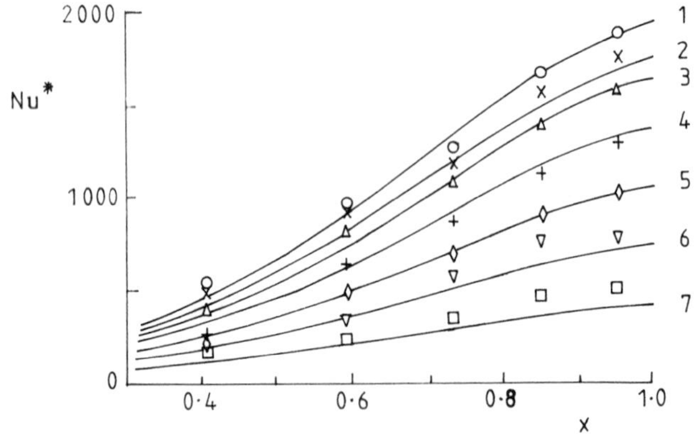

(b)

Figure 5.8 Variation of Nu^* with x for the free disc: $Pr = 0.71$ (Northrop and Owen 1988, Ong 1988)

——— **Numerical solution**

(a) $n \simeq -0.2$

Experiment	○	×	△	+	◊
Curve	1	2	3	4	5
$Re_\phi \times 10^{-6}$	2.65	1.08	0.55	0.27	0.14

(b) $n \simeq 0.6$

Experiment	○	×	△	+	◊	▽	□
Curve	1	2	3	4	5	6	7
$Re_\phi \times 10^{-6}$	1.71	1.59	1.36	1.11	0.83	0.55	0.28

CHAPTER 6
Rotor-Stator Systems with No Superposed Flow

For most of this chapter, it is assumed that one disc (the rotor) of radius b rotates with angular velocity Ω about the z-axis, and that the other (the stator) is stationary. The discs are assumed to be a distance s apart, so it is natural to define a *gap Reynolds number*, Re_*, such that

$$Re_* = \Omega s^2/\nu = G^2 Re_\varphi, \tag{6.01}$$

where $G = s/b$ is the *gap ratio* and Re_φ is the rotational Reynolds number used in the previous chapters. When Re_* is large, it is to be expected that there will be a boundary layer on at least one of the discs, with a region of inviscid flow somewhere between them. When Re_* is small, Couette-type flow is expected in which the viscous region fills the whole space between the discs. Whether Re_* is large or small, the flow may be laminar or turbulent.

It is convenient to follow Daily and Nece (1960) and to refer to four regimes of flow; these are indicated in Figure 6.1. The demarcation lines are obtained by considering the empirical expressions (given in equations (6.08), (6.20), (6.09) and (6.25) for Regimes I, II, III and IV respectively) obtained by Daily and Nece from their measured moment coefficients.

The convention, used throughout the discussion of rotor-stator systems, is to use an asterisk to indicate values of flow quantities when there is no superposed flow

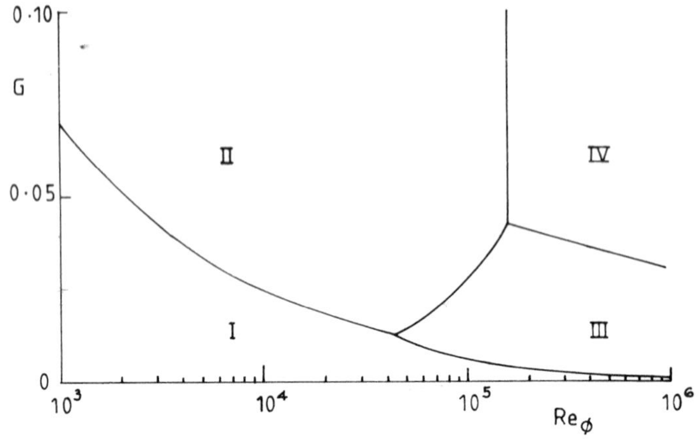

Figure 6.1 The four regimes of flow

Regime	Clearance	Flow	Equation
I	Small	Laminar	(6.08)
II	Large	Laminar	(6.20)
III	Small	Turbulent	(6.09)
IV	Large	Turbulent	(6.25)

through the system. For example β^* and C_M^* indicate the value of β and C_M respectively when there is no superposed flow. This nomenclature should not be confused with the use of an asterisk to denote Nusselt numbers based on the adiabatic-temperature difference (see, for example, equation (2.116)).

6.1 COUETTE FLOW: SMALL CLEARANCE (Regimes I and III)

6.1.1 Theory

When Re_s is sufficiently small, inertial terms are negligible throughout the space between the rotor and the stator. It follows that equation (2.37) reduces to

$$\frac{\partial \tau_\phi}{\partial z} = 0 \qquad (6.02)$$

for both laminar and turbulent flow.

When the flow is laminar, the tangential component of shear stress throughout the boundary layer is

$$\tau_{\phi,\,lam} = \frac{\mu\Omega r}{s} \tag{6.03}$$

and the moment coefficient is

$$C^*_{M,\,lam} = \pi G^{-1} Re_{\phi}^{-1} . \tag{6.04}$$

For turbulent flow in a pipe of diameter d, the shear stress given by Blasius's (1913) law of friction is

$$\frac{\tau_0}{\rho \bar{U}^2} = 0.03955 \left(\frac{\nu}{\bar{U}d}\right)^{\frac{1}{4}} \tag{6.05}$$

where $\bar{U}$ is the bulk-average velocity in the pipe. (This is equivalent to the form given in equation (4.05) with $n = 7$ and $\bar{U} = 0.8U$.) Soo (1958) assumed that this equation is valid for the rotor-stator system with a small clearance if d is replaced by s and $\bar{U}$ by $\frac{1}{2}\Omega r$. This gives

$$\frac{\tau_{\phi,\,turb}}{\rho\Omega^2 b^2} = 0.01176\, G^{-\frac{1}{4}} Re_{\phi}^{-\frac{1}{4}} x^{\frac{7}{4}} , \tag{6.06}$$

and the resulting moment coefficient is

$$C^*_{M,\,turb} = 0.0308\, G^{-\frac{1}{4}} Re_{\phi}^{-\frac{1}{4}} . \tag{6.07}$$

6.1.2 Experiment

Daily and Nece carried out a comprehensive theoretical and experimental study of the flow inside an enclosed rotor-stator system. The experiments were made using a disc of radius 249 mm with a radial clearance of 6 mm and gap ratios G of 0.0127, 0.0255, 0.0637, 0.115 and 0.217. The stationary housing could be filled with either water or lubricating oils of different viscosities, and rotational Reynolds numbers from $Re_{\phi} = 10^3$ to 10^7 were achieved. By using rotating discs of either 6 mm or 12 mm thickness, corrections were made to the measured moment coefficients to account for the effect of disc thickness.

Frictional torque was measured using strain gauges attached to the central shaft of the rotor, and velocity measurements were made using "aerodynamic probes". Pressure distributions were measured by means of static-pressure taps on the stationary casing at radius ratios of $x = 0.306$, 0.510, 0.713 and 0.917 ($x = r/b$). Some of their torque measurements are shown in Figure 6.2 for $G = 0.0255$.

Both Regimes I and III were observed with the smaller clearances (but neither with the larger) and the empirical correlations for the moment coefficients are

$$C_M^* = \pi G^{-1} Re_\phi^{-1} \qquad \text{(Regime I)}, \tag{6.08}$$

$$C_M^* = 0.040\, G^{-\frac{1}{6}} Re_\phi^{-\frac{1}{4}} \qquad \text{(Regime III)}. \tag{6.09}$$

It may be seen that the empirical relation (6.08) is identical with the theoretical expresssion (6.04). The dependence on Re_ϕ shown in equation (6.09) is the same as that in equation (6.07), but the form of the dependence on G is slightly different.

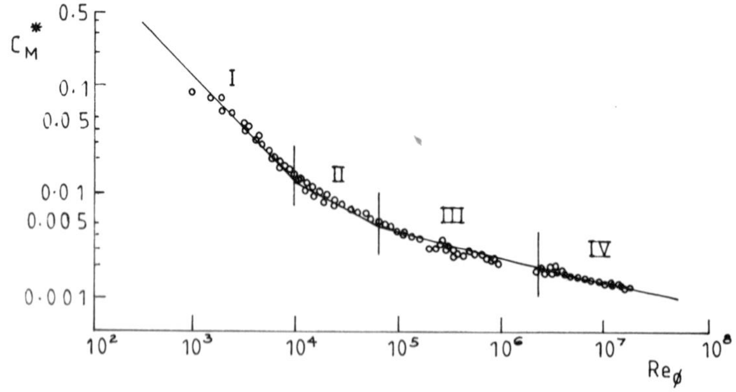

Figure 6.2 Moment coefficient on the rotor: $G = 0.0255$ (Daily and Nece 1960)

——— Correlations ∘ Experiment

6.2 FLOW BETWEEN INFINITE DISCS: LARGE CLEARANCE

It often happens that the ratio G is small even when Re_* is large. Many authors have considered the limiting case in which $G \to 0$ in the hope that it will give a good approximation to what happens in practice; this is equivalent to considering two discs, each of infinite radius, at a finite distance apart.

6.2.1 The Batchelor-Stewartson controversy: an historic paradox

For laminar flow, it is assumed that the velocity of the fluid (v_r, v_ϕ, v_z) is such that $v_r/\Omega r$, $v_\phi/\Omega r$ and v_z are all independent of the radial coordinate r; this was found to be true for all the cases discussed in Chapter 3.

In order to proceed further, it is necessary to assume a suitable model for the flow. Batchelor (1951) argued that there is a core of fluid between the discs rotating with an angular velocity whose magnitude is between 0 and Ω: this implies that there are boundary layers on both discs (see Figure 6.3). The flow on the rotor is analogous to that discussed in Section 3.2.1, with a radial outflow of fluid entrained from the core; the flow on the stator is analogous to that discussed in Section 3.2.2, with a

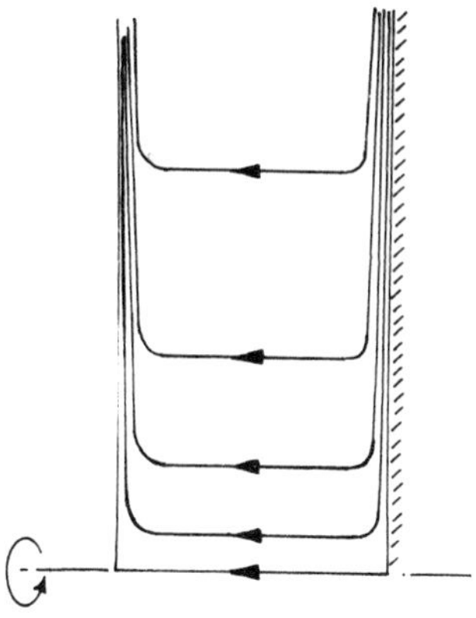

Figure 6.3 Batchelor's model for rotor-stator flow

radial inflow of fluid and an efflux from the boundary layer to the core. Batchelor did not carry out any computations to determine the magnitude of the core rotation.

Stewartson (1953) proposed an alternative model: he suggested that there is a boundary layer on the rotor (analogous to the von Kármán flow on the free disc) in which the tangential component of velocity is reduced from Ωr on the disc to zero far from it; on the stator, such a boundary layer is not necessary. To support this model, he used series solutions to compute the flow for values of Re_s of the order of 10 and found the beginnings of a boundary layer on the rotor and none on the stator. He also conducted experiments with cardboard discs, rotating in free air, and found that the resulting flow did not have a core rotation.

The controversy about Batchelor and Stewartson flows was fuelled by Grohne (1955) who obtained numerical solutions of the von Kármán equations at Reynolds numbers up to $Re_s = 100$. He found that when $Re_s = 10$ no core rotation was apparent, but that when $Re_s = 100$ there was evidence of separate boundary layers and a rotating core.

The first glimmering of light on the problem (at least in retrospect) was given by Picha and Eckert (1958) who made velocity measurements, using aerodynamic probes, between discs of radius 229 mm, values of G from 0.11 to 0.78 and rotor angular speeds between 800 and 3600 rev/min. They found that when the discs were open to the atmosphere no significant core rotation occurred. However, when the discs were surrounded by a stationary casing (or shroud), core rotation did exist and, for a given value of Ω, its angular speed ω increased as G decreased.

Lance and Rogers (1962) extended their own work on a single disc (reported by Rogers and Lance 1960: see Section 3.2.1) to the flow between two discs. For values of $Re_* \leq 100$, their results for the rotor-stator problem showed good agreement with those of Grohne. However, although they showed that Batchelor flow can exist between a rotating and a stationary disc, their results did not disprove the possible existence of Stewartson flow.

Pearson (1965) solved the time-dependent problem in which the rotor is started impulsively from rest. He found that Batchelor flow developed after a sufficient time (about 30 revolutions when $Re_* = 100$, and 150 revolutions when $Re_* = 1000$) giving a flow similar to the steady-state solution found by Lance and Rogers.

The final clue was given by Mellor, Chapple and Stokes (1968) who demonstrated that, in the general case, solutions of the von Kármán equations are not unique. Their numerical results showed that there were multi-cell solutions; in each cell (bounded by planes of constant z in which $v_z = 0$) the fluid was completely self-contained and circulated to and from infinity without crossing the cell boundaries. In particular, for a single cell, their solutions gave two branches separated by a singular point where $Re_* \to \infty$. The first of these branches gives a solution which approaches Batchelor flow as $Re_* \to \infty$; and it is to be presumed that Stewartson's results, obtained at $Re_* = 10$, were from the second branch of the single-cell solution. The work of other earlier authors involved implicit assumptions about the nature of the solution, and each implicitly excluded one or other of the two branches.

Mellor *et al* also made measurements, using a hot-wire anemometer, in an open rotor-stator system in which discs of radius 109 mm were separated by an air gap of 3.18 mm (giving $G = 0.029$). Measurements of the radial and axial components of velocity were made at radial locations of $x = 0.413$ and 0.699 for $Re_* = 50$ and 100. For $Re_* = 100$, there was evidence of a core whose tangential component of velocity was about 30% less than that of the rotor.

It is now clear that the structure of the flow between infinite discs is not unique, and the question arises as to the relevance of the work to finite discs. The experimental work described above has demonstrated that the conditions at the edge of finite discs can affect the type of flow which occurs; these conditions appear to specify the flow in a real rotor-stator problem. Once the appropriate model is determined (that is, the number of cells between the discs and the branch of the solution for a single-cell model), the solution for the infinite disc is unique and may be expected to give a good approximation for a finite disc. For most of the rotor-stator systems found in engineering practice, it is Batchelor-type flow that is appropriate: for an enclosed rotor-stator system, fluid leaving the boundary layer on the rotor moves axially across to the stator in a boundary layer on the cylindrical shroud.

Since the work of Mellor *et al*, many authors have found theoretical evidence of multiple solutions of the von Kármán equations. These solutions, while of considerable academic interest, have little practical importance to the engineer. The interested reader is referred to the comprehensive review of these flows by Zandbergen and Dijkstra (1987).

6.2.2 Theoretical solutions for laminar flow

When Batchelor flow occurs between the rotor and the stator, there is an inviscid core in which the fluid rotates with an angular velocity ω, the value of which is not known *a priori*. The generalized laminar-flow equations (3.35) to (3.37) are valid throughout the whole region with

$$A = \Omega, \quad B = \omega^2 = \beta^2\Omega^2, \tag{6.10}$$

where $\beta = \omega/\Omega$. Since ω (and, therefore, β) is unknown, six boundary conditions are needed instead of the five in

Section 3.2.1. These conditions are

$$F_3(0) = 0, \quad G_3(0) = 1, \quad H_3(0) = 0 \tag{6.11}$$

and

$$F_3(Re_*^{\frac{1}{2}}) = G_3(Re_*^{\frac{1}{2}}) = H_3(Re_*^{\frac{1}{2}}) = 0. \tag{6.12}$$

(F_3, G_3 and H_3 are the functions F, G and H of Section 3.2.1.) Lance and Rogers (1962) solved these equations for a range of values of Re_* up to 441, and their results for a rotor-stator system are shown in Figure 6.4. The tangential component of velocity changed from a linear shear when $Re_* \simeq 0$ to a flow with a rotating core when $Re_* = 441$. In the latter case, the angular speed of the core was approximately $0.3\,\Omega$.

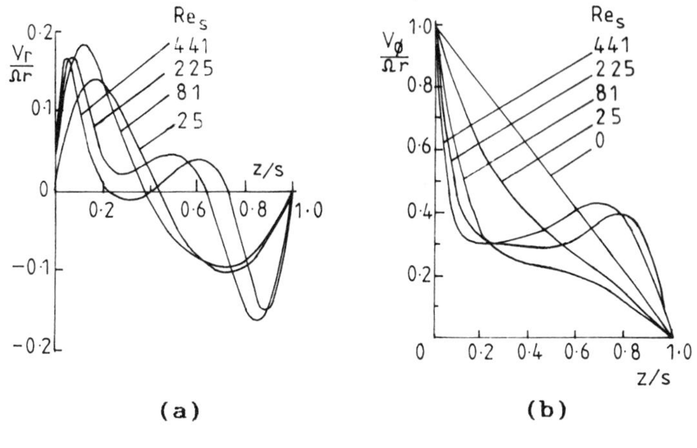

Figure 6.4 Effect of Re_* on laminar velocity profiles (Lance and Rogers 1962)

As long as Re_* is sufficiently large, and Batchelor flow exists with an inviscid core rotation ω, it is useful to consider the system in two parts: the rotor and the stator are thought of as single discs which (if ω is constant) are similar to those discussed respectively in Sections

3.2.1 and 3.2.2. Using equation (3.26), it can be seen that the mass flow-rate on the rotor is $\dot{m}_0$ and that on the stator is $\dot{m}_*$, where

$$\dot{m}_0 = -\rho\pi r^2(\Omega\nu)^{\frac{1}{2}}H_{\infty,0}, \quad \dot{m}_* = -\rho\pi r^2(\beta\Omega\nu)^{\frac{1}{2}}H_{\infty,*}; \tag{6.13}$$

here $H_{\infty,0}$ is given in Table 3.2 as a function of β, and from equation (3.67) $H_{\infty,*} = 1.3494$. Since there is no superposed flow, it follows that

$$\dot{m}_0 + \dot{m}_* = 0. \tag{6.14}$$

If linear interpolation is used in Table 3.2, it is easy to see that $\beta^* \simeq 0.314$, which is in reasonable agreement with the value of 0.313 quoted by Dijkstra and van Heijst (1983) for infinite discs.

Using the Ekman-layer approximation, equations (3.63) and (3.74) show that both $H_{\infty,0}$ and $H_{\infty,*}$ are approximately unity. It follows from equations (6.13) and (6.14) that

$$(1 - \beta^*) - \beta^{*\frac{1}{2}} = 0. \tag{6.15}$$

This is a quadratic equation for $\beta^{*\frac{1}{2}}$ whose solution is $\beta^* = 0.382$, which is about 25% higher than the value 0.313 obtained by Dijkstra and van Heijst using the full equations. The numerical and experimental results of these authors are shown, together with the two approximate solutions, in Figure 6.5; these results are discussed in more detail in Section 6.3.

6.2.3 Theoretical solutions for turbulent flow

Assuming that there is an inviscid core rotating with a constant angular velocity $\omega = \beta\Omega$, it is possible (as for laminar flow) to treat the rotor and the stator separately. If $\frac{1}{7}$-power-law profiles are used, $\dot{m}_0$ and $\dot{m}_*$ can be obtained from equations (4.69), (4.70) and (4.81), so that

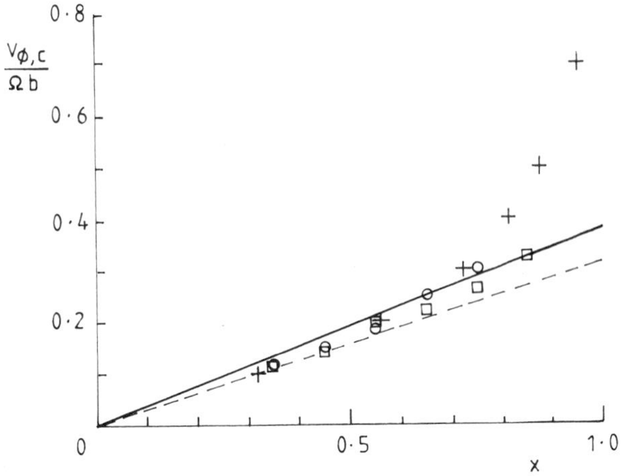

Figure 6.5 Variation of $v_{\phi,c}/\Omega b$ with x: comparison between Ekman-layer approximation and the results of Dijkstra and van Heijst (1983) for $G = 0.07$

Theory:	———	Ekman-layer approximation ($\beta^* = 0.382$)
	- - - - -	infinite-disc solution ($\beta^* = 0.313$)
	+	numerical solution, rotating shroud
Experiment:	○	rotating shroud
	□	stationary shroud

$$\frac{\dot{m}_0}{\mu r} = \epsilon_{m,0}(x^2 Re_\phi)^{\frac{4}{5}} = \frac{49\pi}{60}\,\mathrm{sgn}(1-\beta^*)\alpha_0\gamma_0|1-\beta^*|^{\frac{8}{5}}(x^2 Re_\phi)^{\frac{4}{5}}, \quad (6.16)$$

where $\alpha_0 = \alpha_0(\beta^*)$, $\gamma_0 = \gamma(\beta^*)$; $\alpha_0(\beta)$ and $\gamma_0(\beta)$ are given by equations (4.67) and (4.68) respectively. Similarly, $\dot{m}_s$ is given by equation (4.81) with $Re'_\phi = \beta^* Re_\phi$, so that

$$\frac{\dot{m}_s}{\mu r} = -\frac{49\pi}{60}\alpha_s\gamma_s\beta^{*\frac{4}{5}}(x^2 Re_\phi)^{\frac{4}{5}} \quad (6.17)$$

where $\alpha_s = \alpha_s(\beta^*)$, $\gamma_s = \gamma_s(\beta^*)$; $\alpha_s(\beta)$ and $\gamma_s(\beta)$ satisfy equations (4.82) and (4.83) respectively. Equation (6.14) then becomes

$$\frac{49\pi}{60}(x^2 Re_\phi)^{\frac{4}{5}}[\alpha_0\gamma_0\,\mathrm{sgn}(1-\beta^*)|1-\beta|^{\frac{8}{5}} - \alpha_s\gamma_s\beta^{*\frac{4}{5}}] = 0. \quad (6.18)$$

In the Ekman-layer approximation, α_0 and α_s are both approximated by $\alpha_0(1) = 0.553$, and γ_0 and γ_s by $\gamma_0(1) = 0.0983$. Equation (6.18) then reduces to

$$\mathrm{sgn}(1 - \beta^*)\,|1 - \beta^*|^{\frac{8}{5}} - \beta^{*\frac{4}{5}} = 0. \qquad (6.19)$$

This implicit equation for β^* can be solved to give, as for laminar flow, $\beta^* = 0.382$.

(If the exact expressions for α_0, α_s, γ_0 and γ_s are used in equation (6.14), there is no solution of the equation with β^* constant; if β^* is not constant, there is a violation of the assumptions made in the derivations in Section 4.2.3. A solution has, in fact, been derived assuming that the exact expressions derived for constant β are valid locally: this assumption is justified when the Ekman-layer approximation is made since, in this case, the equations themselves are valid locally. The method is not presented here, however, since the results do not agree with either the numerical predictions of Morse (see Section 6.4.1) or the experimental results. It seems, therefore, that this approach cannot be used with the full equations if $\frac{1}{7}$-power-law profiles are used on both discs.)

Owen (1988) suggested an "improved approximation" in which the full equations are used for the rotor and the Ekman-layer equations for the stator. This is equivalent to using equations (4.67) and (4.68) for $\alpha_0(\beta^*)$, with $\alpha_s = \alpha_0(1) = 0.553$ and $\gamma_s = \gamma_0(1) = 0.0983$ as above. Equation (6.18) is then solved implicitly to give $\beta^* = 0.426$. The "improved approximation" can be simplified by using ϵ_m', given by equation (4.75) for $\beta < 1$, for the rotor instead of ϵ_m in equation (4.69). This results in the equation

$$0.154\,\beta^{*\frac{8}{5}} - 0.526\,\beta^{*\frac{4}{5}} + 0.228 = 0 \qquad (6.20)$$

which is a quadratic in $\beta^{*\frac{4}{5}}$ whose solution gives $\beta^* = 0.431$. The above simplification will also be used in Section 7.3.2, where a superposed flow is discussed.

6.3 ENCLOSED SYSTEMS: LARGE CLEARANCE, LAMINAR FLOW (Regime II)

6.3.1 Theory

Dijkstra and van Heijst (1983) used a finite-difference technique to obtain numerical solutions of the Navier-Stokes equations for a rotor-stator system, with a shroud which could be rotating or stationary. Resolution near the wall was improved using "coordinate-stretching functions"; for the majority of the computations, a 20x20 grid was uniformly distributed in the transformed plane. Their results are shown in Figure 6.5; for x less than about 0.5, the numerical solution is in good agreement with their solution for infinite discs ($\beta^* = 0.313$). For larger values of x, the presence of the shroud causes significant departures from the solution for infinite discs (in which the angular velocity of the core is independent of x); this invalidates the assumption made by many authors that ω depends not on x but only on the gap ratio G.

Vaughan (1986) also obtained finite-difference solutions of the Navier-Stokes equations for a rotor-stator system. His results for $G = 0.1$ and $Re_\phi = 5\text{x}10^4$ are shown in Figure 6.6 (at $x = 0.5$ for $v_r/\Omega r$ and at $x = 0.6$ for $v_\phi/\Omega r$). The computations were carried out using a grid of 59x65 points with a fine mesh near the walls; grid-independency tests suggested that truncation errors were negligible.

A number of authors have used techniques involving the momentum-integral equations to estimate the flow in a rotor-stator system: these include Schultz-Grunow (1935) who considered infinite discs, Okaya and Hasegawa (1939) who modified the solution to take account of differences in radius between the rotor and the stator, and Daily and Nece (1960) who considered the effect of the cylindrical casing on the moment coefficient. However, these authors

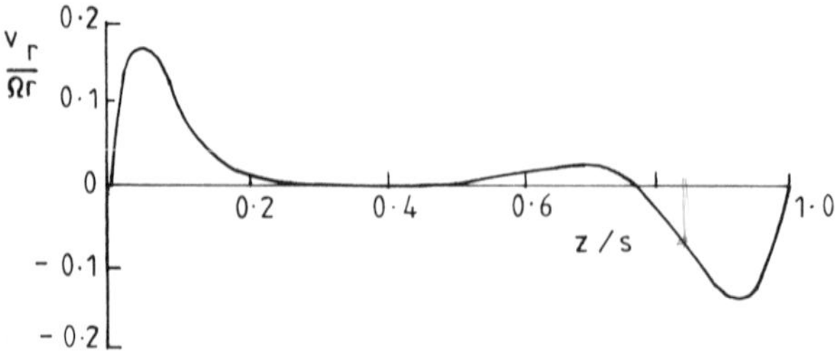

(a) radial velocity: $x = 0.5$

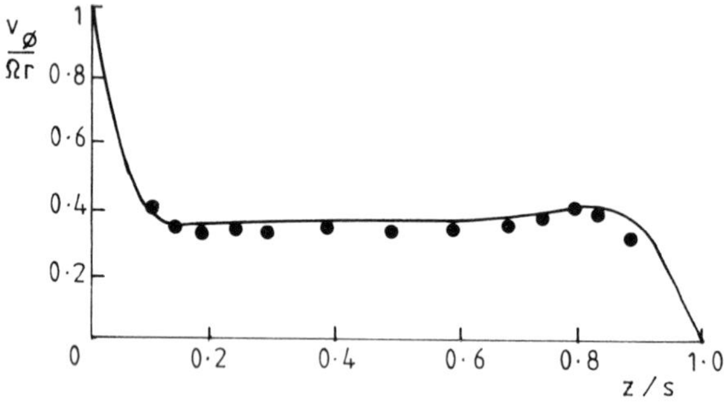

(b) tangential velocity: $x = 0.6$

Figure 6.6 Axial variation of laminar velocity profiles: $G = 0.1$, $Re_\phi = 5\text{x}10^4$

——— numerical predictions (Vaughan 1986)

● experimental data (Sambo 1983)

---

assumed, explicitly or implicitly, that the axial component of velocity was zero in the inviscid core. In consequence the integral equations were incomplete and the equation of continuity was not satisfied. These solutions are not presented here, and the interested reader is referred to the original papers for details.

6.3.2 Experiment

Measurements of velocity distributions and moment coefficients inside rotating rotor-stator systems have been made by many workers: these include Schultz-Grunow

(1935), Pantell (1950), Sedach (1957), and Daily and Nece (1960) whose work is described here in more detail.

In their experiments (see Section 6.1.2), Daily and Nece found that Regimes II and IV occurred for the larger gap ratios. Their torque measurements showed that

$$C_M^* = 1.85\, G^{\frac{1}{10}} Re_\phi^{-\frac{1}{2}} \quad \text{(Regime II)}. \tag{6.21}$$

In contrast to their expression for Regime I (see equation (6.08)), the moment on the rotor increases as G increases. This is caused by the extra shear imposed by the shroud: the moment on the rotor is balanced by the sum of the moments on the stator and on the shroud. The variation of C_M^* with Re_ϕ is shown in Figure 6.2 for $G = 0.0255$.

When G is small, equation (6.08) shows that, for a constant value of Re_ϕ, the moment coefficient decreases as G increases. If the clearance is increased until the flow has changed from Regime I to Regime II, equation (6.21) shows that the moment coefficient increases as G increases. The coefficient takes a minimum value, $C_{M,\min}$, when the expressions in equations (6.08) and (6.21) are equal: this happens when $G = G_{crit}$, where

$$G_{crit} = 1.62\, Re_\phi^{-\frac{5}{11}}, \tag{6.22}$$

and

$$C_{M,min}^* = 1.94\, Re_\phi^{-\frac{6}{11}}. \tag{6.23}$$

(The relationship given in equation (6.22) defines the curve between Regimes I and II in Figure 6.1.) If the moment coefficient for the free disc, $C_{M,\mathrm{fd}}$, is given by equation (3.30), it is clear that

$$\frac{C_{M,min}^*}{C_{M,fd}} \simeq Re_\phi^{-\frac{1}{22}}. \tag{6.24}$$

It follows that the moment coefficient for the rotor can be significantly less than the free-disc value. For example, when $Re_\phi = 10^3$ the ratio $C_{M,\min}^*/C_{M,\mathrm{fd}}$ is 0.73 and when $Re_\varphi = 1.5\text{x}10^5$ the ratio is 0.58 (corresponding to $G_{crit} = 0.069$ and 0.0072 respectively).

Measurements of the angular velocity in the mid-axial plane of the cavity were made by Daily and Nece at $x = 0.765$, for a rotational Reynolds number $Re_\varphi = 4.2\text{x}10^4$ and for three different clearance ratios of $G = 0.051$, 0.102 and 0.217; in each case the flow was of Regime II type. The measured values of β^* for the three clearances were 0.46, 0.44 and 0.36 respectively: these are somewhat greater than the value given for infinite discs ($\beta^* = 0.313$) by Dijkstra and van Heijst (1983).

If it is assumed that equation (2.29) is valid in the core, integration with respect to r gives

$$\frac{p_2 - p_1}{\frac{1}{2}\rho\Omega^2 b^2} = \beta^{*2}(x_2^2 - x_1^2), \tag{6.25}$$

where p_1, p_2 are the static pressures when $x = x_1$, x_2 respectively. The pressure distributions measured by Daily and Nece were consistent with the form given in equation (6.25), but the pressure differences were smaller than those predicted using their measured values of β^*. For example, when $G = 0.0637$ and 0.102 the measured pressure differences were only about 70% of the values predicted. (The measurements were consistent with $\beta^* = 0.37$ but, as noted below, β^* is not constant in an enclosed rotor-stator system.)

Dijkstra and van Heijst (1983) made velocity measurements in the water between two horizontal discs, to the lower of which was attached a cylindrical shroud. The diameters of the upper and lower discs were 1000 and 900 mm respectively and the axial clearance was 35 mm (so that $G = 0.07$); either disc could be rotated separately. Velocities were determined using stereo-photography: polystyrene particles in the water were illuminated by flashes of stroboscopic light, and the resulting traces were photographed. The dependence on x of their measured tangential velocities when $Re_\phi = 2\times10^5$ is shown in Figure 6.5. Results for the shroud rotating and stationary are shown; they differ only slightly from each other and from the infinite-disc predictions at small and moderate radii.

Agreement between the experiments of Dijkstra and van Heijst and their numerical computations (carried out for a rotating shroud) are good at all radii. The solution (6.15) obtained by using the Ekman-layer approximation is also shown in the figure; it may be seen that (fortuitously but usefully!) the value $\beta^* = 0.382$ agrees with the experiments even better than the more exact solution ($\beta^* = 0.313$) for infinite discs. It should be noted that, for finite discs, the core rotation is not independent of radius: except very close to the shroud, β^* increases as x increases.

Sambo (1983) used laser-doppler anemometry (LDA) to measure the velocity of air in a rotor-stator system with discs of diameter 381 mm spaced 19 mm apart; the angular speed of the rotor was up to 4100 rev/min. Some of his results are shown in Figure 6.6, and agreement with Vaughan's theoretical predictions is good.

Szeri, Schneider, Labbe and Kaufmann (1983) used laser-doppler anemometry to make velocity measurements in the water between two discs rotating with different angular speeds. For the case in which one disc was stationary and the other rotating, Batchelor-type flow was observed.

An interesting application of the enclosed rotor-stator system was considered by Reshotko and Rosenthal (1968). The authors analyzed "slinger seals", an illustration of which is shown in Figure 6.7. Oil, or some other liquid, is used to prevent the flow of air, or some other gas, from one side of the system to the other. The oil is centrifuged outwards, and the differential air pressure is balanced by the centripetal acceleration of the rotating oil film. Using laminar infinite-disc theory, the authors

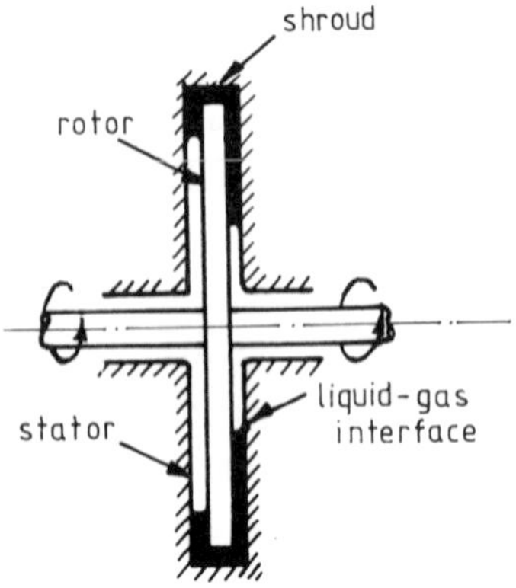

Figure 6.7
Slinger seal (Reshotko and Rosenthal 1968)

calculated the ratio of the frictional moment to the pressure differences. This ratio was used to define a figure-of-merit which suggested that, for laminar flow, the minimum power per unit pressure difference occurs for $3 < Re_z < 12$.

6.4 ENCLOSED SYSTEMS: LARGE CLEARANCE, TURBULENT FLOW (Regime IV)

6.4.1 Theory

Cooper and Reshotko (1975) used a technique which is intermediate between the numerical solution of the elliptic equations and a momentum-integral technique. Their method involved the use of stream functions that, in effect, converted the full boundary-layer equations from partial to ordinary differential equations at each radial location. The solutions for the rotor and stator were matched in the core by the used of a "shooting" technique.

They used an effective-viscosity model based on the two-layer model of Cebeci and Smith (1974), and transition from laminar to turbulent flow was assumed to occur in the range $1.6\text{x}10^5 < x^2 Re_\phi < 2.5\text{x}10^5$. Their computations were carried out for $Re_z = 296$ and 2852, and for $0 \le Re_\phi \le 10^7$; this includes all the four regimes of flow shown in Figure 6.1. They found expressions for C_M which were similar to those given in equations (6.08), (6.09), (6.20) and (6.25), found experimentally by Daily and Nece.

Vaughan (1986) used the technique mentioned in Section 6.3.1 to solve the elliptic equations for turbulent flow. He used a mixing-length model (which was a modified form of that proposed by Koosinlin, Launder and Sharma 1974), and a non-uniform grid with 70x90 nodes. His computed velocity profiles for $x = 0.765$, $G = 0.0637$ and $Re_{\phi} = 4.4 \times 10^6$ are shown in Figure 6.8. It is interesting to note that, as for laminar flow, the computed value of v_{ϕ} shows a maximum near the stator and a minimum near the rotor.

Morse (1989) obtained numerical solutions of the elliptic equations, using an anisotropic low-Reynolds number $k-\epsilon$ turbulence model. (He ensured that transition from laminar to turbulent flow occurred at $Re_{\phi} \simeq 3 \times 10^5$ by "enhancing" the energy-production term in the turbulence model: 0.2% of the "turbulent viscosity", determined from a simple mixing-length model, was added to the production term.) The radial variations of β^*, shown in Figure 6.9, were obtained using a finite-difference grid with 65 axial and 115 radial nodes. The gap ratio was $G = 0.1$, an axis of symmetry was assumed at $x = 0$ and a stationary shroud at $x = 1$.

At $Re_{\phi} = 10^5$, Figure 6.9 shows that the flow remains laminar and $\beta^* = 0.315$ at $x = 0$: this is very close to the value of 0.313 given in Section 6.3 for infinite discs. Transition occurs on the rotor at $Re_{\phi} \simeq 3.5 \times 10^5$, and the turbulence is amplified in the highly sheared flow adjacent to the shroud. This results in a greater turbulence level on the stator than on the rotor as the flow moves radially inwards, and, for $Re_{\phi} = 4 \times 10^5$, β^* falls below the laminar levels! However, for $Re_{\phi} > 6 \times 10^5$, transition occurs at increasingly smaller radii on the

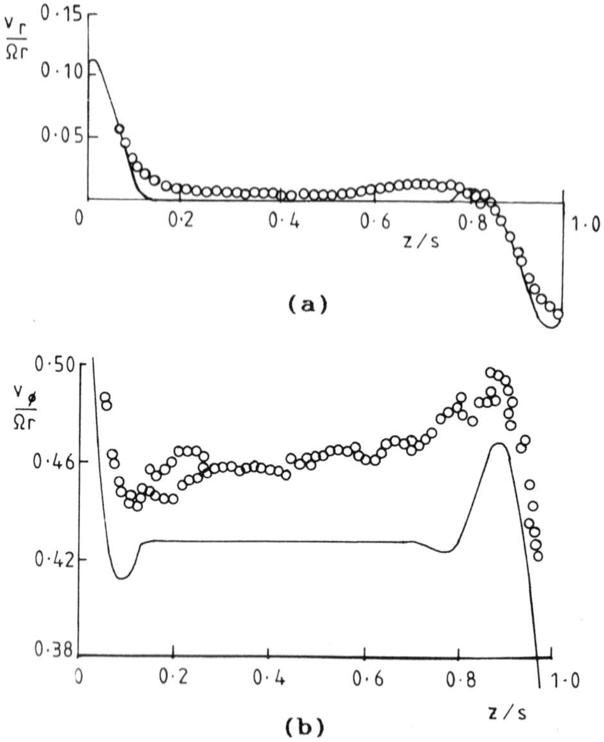

Figure 6.8 Axial variation of turbulent velocity profiles: $G = 0.064$, $Re_\phi = 4.4\text{x}10^6$, $x = 0.765$

———— numerical predictions (Vaughan 1986)

o experimental data (Daily and Nece 1960)

---

rotor, and β^* increases as Re_ϕ increases. It can also be seen from Figure 6.9 that there is a "plateau region" where $\beta^* \simeq 0.43$, which is the value given by the "improved approximation" obtained from equation (6.20). The radial extent of this plateau region increases as Re_ϕ increases and, at $Re_\phi = 10^7$, $\beta^* \simeq 0.43$ for $0.2 \lesssim x \lesssim 0.9$. It is also interesting to note that, for both laminar and turbulent flow, the *stationary* shroud causes an increase in β^* near $x = 1$.

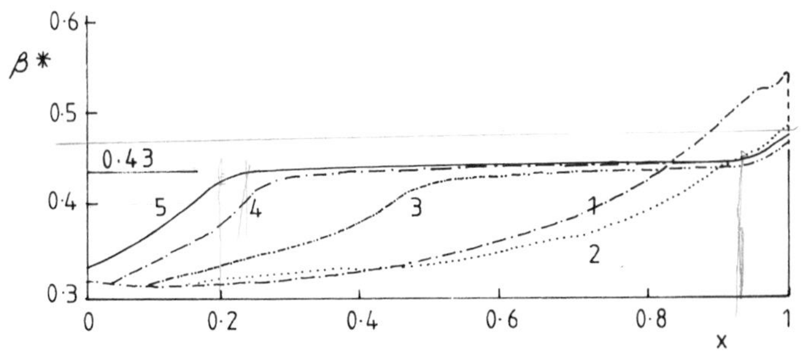

Figure 6.9 Effect of Re_ϕ on radial variation of β^*: $G = 0.1$ (Morse 1989)

Curve	1	2	3	4	5
Re_ϕ	10^5	4×10^5	10^6	4×10^6	10^7

Ketola and McGrew (1968), whose research (like that of Reshotko and Rosenthal 1968 discussed in Section 6.3.2) was concerned with "slinger seals", extended the work of Daily and Nece using (incomplete) momentum-integral equations to determine the moment coefficient for the rotor. They considered the case where only part of the rotating disc was "wetted": this is equivalent to integrating the moment from $r = a'$ to $r = b'$, where a' and b' represent respectively the inner and outer radii of the liquid film.

6.4.2 Experiment

Daily and Nece (1960), using the apparatus described above, found that

$$C_M^* = 0.0510\, G^{\frac{1}{10}} Re_\phi^{-\frac{1}{5}} \qquad \text{(Regime IV).} \qquad (6.26)$$

(This gives a value of C_M^* within 2% of that computed for $G = 0.0637$ and $Re_\phi = 4.4\times10^6$ by Vaughan 1986.)

Comparison of equations (6.09) and (6.26) shows that, as for laminar flow, C_M^* takes a minimum value $C^*_{M,min}$ when $G = G_{crit}$ where

$$G_{crit} = 0.402\, Re_{\phi}^{-\frac{3}{16}}, \tag{6.27}$$

$$C^{*}_{M,min} = 0.0466\, Re_{\phi}^{-\frac{7}{32}}. \tag{6.28}$$

If $C_{M,fd}$ is the free-disc value based on equation (4.18), it follows that

$$\frac{C^{*}_{M,min}}{C_{M,fd}} = 0.639 Re_{\phi}^{-\frac{3}{160}}. \tag{6.29}$$

As Re_{ϕ} is increased from $1.5\text{x}10^5$ to 10^7, the value of $C^{*}_{M,min}/C_{M,fd}$ decreases from 0.51 (with $G_{crit} = 0.043$) to 0.47 (with $G_{crit} = 0.020$). It follows that, if the gap ratio is chosen to be G_{crit}, the moment coefficient for an enclosed disc can be reduced to approximately half that of the free disc.

For $G = 0.0637$, 0.102 and 0.217, the values of β^* measured by Daily and Nece in Regime IV at $x = 0.765$ were 0.460, 0.44 and 0.412 respectively. However, when used in equation (6.24), these values of β^* give overestimates of their measured pressure gradients by approximately 10% at the two smaller values of G. In fact, for $G = 0.0637$ and 0.102, the pressure measurements were best fitted by a value of $\beta^* \simeq 0.43$, which is in good agreement with the approximate solution obtained from equation (6.20) and with the computed value obtained by Morse.

The values of v_r and v_{ϕ} measured by Daily and Nece (for $x = 0.765$, $G = 0.0637$ and $Re_{\phi} = 4.4\text{x}10^6$) are shown in Figure 6.8 where they can be compared with the numerical solutions of Vaughan. The agreement between the computed and measured values of v_r is good. As stated above, the pressure distribution suggests that the measured values of v_{ϕ} may be high in the core region; the scale chosen in Figure 6.8b exaggerates the difference between the measured and computed values. Daily, Ernst and Asbedian (1964), using apparatus similar to that of Daily and Nece, obtained the results shown in Table 6.1.

Table 6.1 Variation of β^* and $C_M^* Re_\phi^{\frac{1}{5}}$ with G (Daily *et al* 1964)

G	0.0276	0.069	0.124
β^*	0.475	0.45	0.42
$C_M^* Re_\phi^{\frac{1}{5}}$	0.0367	0.0397	0.0420

Cooper and Reshotko (1975) carried out experiments using a test rig with a disc of 203 mm diameter rotating up to 13000 rev/min inside a stationary enclosure. The moment on the rotating disc was determined from the measured torque necessary to prevent the enclosure, which was mounted on ball bearings, from rotating. Their measured moment coefficients were consistent with those of Daily and Nece, and agreement with their own computed results was good.

6.5 FLOW INSTABILITIES AND VORTEX BREAKDOWN

6.5.1 Flow instabilities

A jet of fluid flowing in a quiescent environment entrains fluid from its surroundings. If a stationary plate is placed close to one side of the jet, and parallel to the jet axis, entrainment is restricted on that side creating a low pressure region. The jet is then deflected towards the plate until it becomes attached: a *wall jet* is formed. In an open rotor-stator system, the radial jet of fluid leaving the rotor can, if the stator is large enough and close enough, "snap on" to the stator. This restricts the inflow (or ingress) necessary to satisfy continuity, and the resulting flow becomes nonaxisymmetric and/or unsteady.

Maroti, Deak and Kreith (1960) studied the unsteady flow that occurs when large stationary surfaces are placed on either side of a rotating disc. The two stators were of larger diameter than the disc and were attached to each other by a series of uniformly-spaced bolts. For the case of a disc of 406 mm diameter and 12.7 mm thickness rotating between stators of 610 mm diameter, the authors observed that the flow was periodic. There were several distinct regions of inflow and outflow which rotated at a speed slower than that of the disc, and the number of these regions increased when the rotational speed was increased and when the axial spacing was decreased.

Instabilities were also observed by Richardson and Saunders (1963) who plotted the isotherms for a disc of 457 mm diameter and 29 mm thickness rotating at an axial distance of 25 mm from a large stationary surface. The jet from the disc attached itself to the stator, part flowing radially outward and part inward. When large stationary surfaces were placed on either side of the disc, as shown in Figure 6.10, the jet attached itself alternately to each surface creating a periodic instability. These peripheral instabilities can significantly affect the flow and heat transfer near the surface of the disc. Similar effects can occur in shrouded rotor-stator systems with superposed flow, as discussed in Chapters 8 and 9.

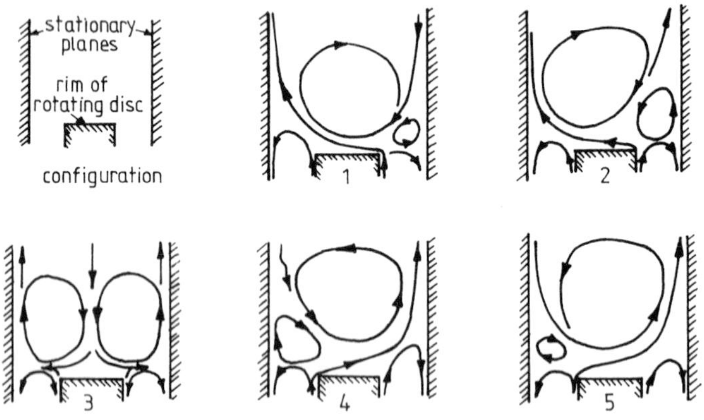

Figure 6.10 Sequence of flow patterns near periphery of rotor (Richardson and Saunders 1963)

6.5.2 Vortex breakdown

Vortex breakdown is an abrupt change in the structure of an axial jet of swirling fluid. Most observations of the phenomenon have been for stationary systems in which swirl is introduced by means of guide vanes, but it can also occur in rotating flows. There is no general agreement about the criteria necessary for the occurrence of vortex breakdown and the topic is controversial: a comprehensive review of the subject is given by Escudier (1988).

Escudier (1984) observed vortex breakdown in a sealed rotor-stator system, filled with a glycerine-water mixture, with large gap ratios ($G > 1$). For the experimental rig, the discs were of 190 mm diameter, and the axial spacing could be varied up to 350 mm. Fluorescent dye was injected into the glycerine mixture at the centre of the stator, and the flow was illuminated by oscillating the beam of a 5W argon ion laser in a diametrical plane through the stationary perspex cylinder that enclosed the rotating and stationary discs.

Flow visualization for $Re_\phi \lesssim 3500$ revealed that axisymmetric vortex breakdown (or "bubble breakdown") could occur along the central (vertical) axis as shown in Figure 6.11 for $G = 2$ and $1002 \leq Re_\phi \leq 1854$. At $Re_\phi = 1002$, there was radial outflow on the rotor (at the bottom of the photographs) and inflow on the stator (at the top), with axial flow from the stator to the rotor at the centre and from the rotor to the stator along the cylindrical side walls. At $Re_\phi = 1449$, the flow came close to stagnation on the centre-line; at $Re_\phi = 1492$ there was a well-defined vortex breakdown with stagnation points on the centre-line upstream and downstream of a bubble of "near-stagnant recirculating fluid". At $Re_\phi = 1854$, there was evidence of an incipient second breakdown below the first one.

For $G < 1.95$, the vortex breakdown manifested itself as a

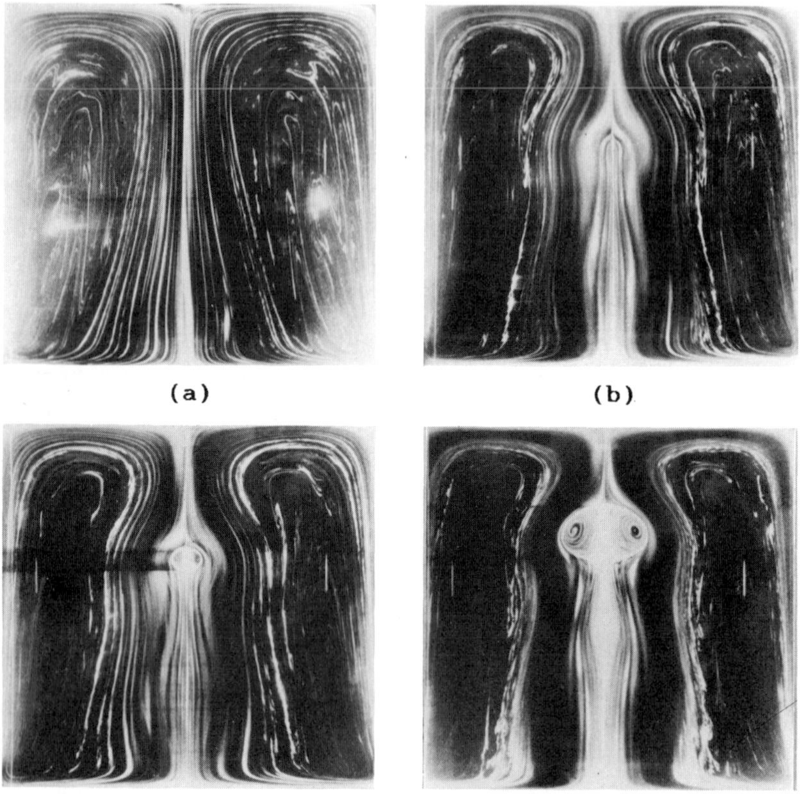

(a) (b) (c) (d)

Figure 6.11 Development of vortex breakdown at centre of an enclosed rotor-stator system: $G = 2$ (Escudier 1984)

Note: stator at top and rotor at bottom of each photograph

single separation bubble; for $1.95 < G < 3.25$, two bubbles could form; for $G = 3.25$ and $Re_\varphi = 2800$, three bubbles were observed. Figure 6.12 shows the delineation of the regimes of vortex breakdown in the G-Re_φ plane. The broken curve with triangular symbols represents the boundary above which the stagnation point of the bubble (the upstream

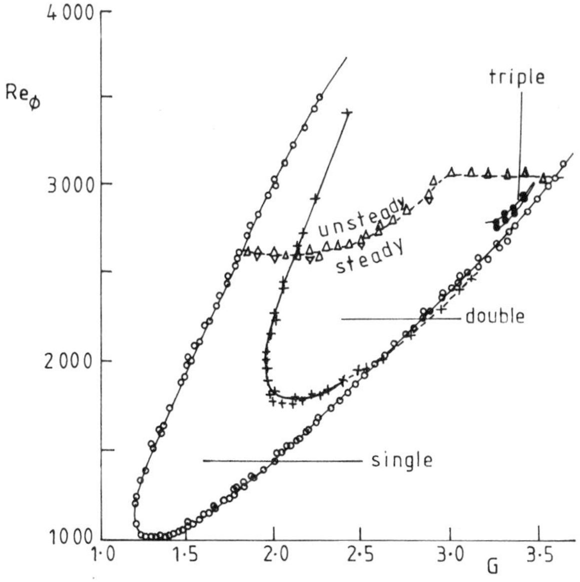

Figure 6.12 Stability boundaries for single, double and triple vortex breakdowns (Escudier 1984)

---

bubble for multiple bubbles) began to oscillate: the oscillations were periodic and axial. As Re_ϕ increased, the motion became disturbed and eventually turbulent, and for $G > 3.1$ the first sign of unsteady motion was a precession of the lower breakdown structure.

In an inviscid rotating fluid, diverging streamlines create an adverse pressure gradient which can cause a stagnation point on the axis of rotation (see Batchelor 1967). According to Lugt and Abboud (1987) it is this inviscid phenomenon (rather than instability in the flow) that creates the stable axisymmetric separation bubbles observed by Escudier. Lugt and Abboud obtained numerical solutions of the axisymmetric Navier-Stokes equations, using a finite-difference scheme, and produced "streamlines" in the rz-plane that were in good agreement with Escudier's photographs of the separation bubbles. They suggested that the enclosed system enforced axisymmetry; however, as discussed in Section 8.2.3, nonaxisymmetric instabilities can occur in open rotor-stator systems with a superposed flow.

6.6 EFFECT OF ROUGHNESS

Nece and Daily (1960) used the apparatus described in Section 6.1.2 to determine the effects of roughness on enclosed rotor-stator systems. Commercial grit papers were bonded to both the rotor and stator, and tests were conducted for relative roughnesses of $k_s/b = 10^{-3}$, $5\text{x}10^{-4}$ and $3.2\text{x}10^{-4}$ (where k_s is the effective height of the roughness as discussed in Section 4.4), for $G = 0.0227$, 0.0609 and 0.112, and for $4\text{x}10^3 < Re_\phi < 6\text{x}10^6$.

The four flow regimes found in their tests with smooth discs occurred also when roughness was present. It was found that roughness had no significant effect on the moment coefficient in the laminar Regimes I and II; for $Re_\phi \gtrsim 2\text{x}10^5$, in the turbulent Regimes III and IV, C_M increased as the roughness increased. Figure 6.13 shows a summary of the results obtained for the variation of C_M with Re_ϕ; and it can be seen that, at large values of Re_ϕ, the moment coefficient becomes independent of the rotational Reynolds number. This is equivalent to the *completely rough* regime discussed in Section 4.4.

From results obtained at the largest values of Re_ϕ, Nece and Daily obtained the correlation

$$C_M^{*\,-\frac{1}{2}} = -5.37 \log_{10}\left(\frac{k_s}{b}\right) - 3.4\,G^{\frac{1}{4}}, \qquad (6.30)$$

which was in good agreement with their nine data points. Although velocity measurements conducted at $G = 0.0227$ showed that merged boundary layers occurred, there were no significant differences of C_M between Regimes III and IV. The authors concluded that roughness effects started to

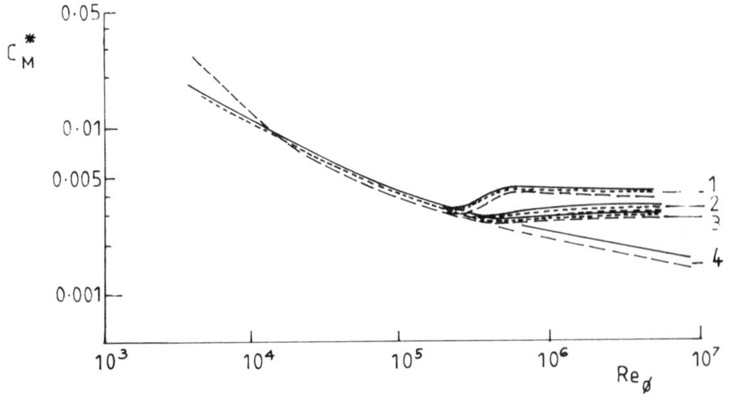

Figure 6.13 Effect of roughness on variation of C_M^* with Re_ϕ (Nece and Daily 1960)

	Group of curves	1	2	3	4
	k_s/b	10^{-3}	$5\text{x}10^{-4}$	$3.2\text{x}10^{-4}$	0
————————	G	0.118	0.118	0.118	0.115
- - - - - - -	G	0.0609	0.0609	0.0609	
— — — —	G	0.0227	0.0227	0.0227	0.0255

become significant for

$$Re_\phi C_M^{*\,-\frac{1}{2}} \simeq 1560\left(\frac{k_s}{b}\right)^{-\frac{2}{5}} \tag{6.31}$$

and that the completely rough regime started at

$$Re_\phi C_M^{*\,-\frac{1}{2}} \simeq 16000\left(\frac{k_s}{b}\right)^{-\frac{1}{10}}. \tag{6.32}$$

Kurokawa, Toyokura, Shinjo and Matsuo (1978) carried out a combined theoretical and experimental study for the case where the rotor and stator could be made independently rough or smooth. The theoretical analysis was similar to that discussed in Section 4.4: the momentum-integral equations for both discs were solved using logarithmic velocity profiles. In their experiments, a disc of 328 mm diameter was rotated in a water-filled casing with radial and axial clearances of 2 mm and 12.8 mm ($G = 0.078$) respectively. The surfaces of the rotor and the stator could be roughened using commercial grit papers with $1.34 \times 10^{-4} < k_s/b < 1.22 \times 10^{-3}$, and tests were conducted at $Re_\phi = 3.1 \times 10^6$.

Tangential velocity profiles, measured with an "aerodynamic probe" at $x = 0.79$, are shown in Figure 6.14 for various combinations of rough and smooth discs. It can be seen that roughness has a significant effect on the tangential component of velocity in the core. For the case where the rotor and the stator are both smooth or both equally rough, $\beta^* \simeq 0.43$ (which agrees with the result discussed in Section 6.4). For different roughnesses on the rotor and the stator, β^* increases as the roughness increases on the rotor or as the roughness decreases on the stator.

For flow over a flat plate with a free-stream velocity U_∞, the *admissible roughness* k_{adm} (that is, the roughness below which the surface is *hydraulically smooth* - see Section 4.4) is given by

$$U_\infty k_{adm}/\nu \simeq 100. \tag{6.33}$$

From their pressure measurements, Kurokawa *et al* concluded that this criterion is valid for a rotor-stator system if U_∞ is replaced by the relative core velocity $\Omega r(1 - \beta^*)$. Assuming that $\beta^* = 0.43$, equation (6.33) becomes

$$\left(\frac{k_{adm}}{b}\right) \simeq 180\,(xRe_\phi)^{-1}. \tag{6.34}$$

It should be noted that this criterion implies that roughness first becomes significant at the outer radii.

Although the authors did not measure moment coefficients for the roughened disc, their solutions of the integral equations were in good agreement with the measurements of Nece and Daily. Also the values of Re_ϕ calculated from equation (6.34), for $x = 1$, are consistent with the values where the curves for smooth and rough discs diverge in Figure 6.13.

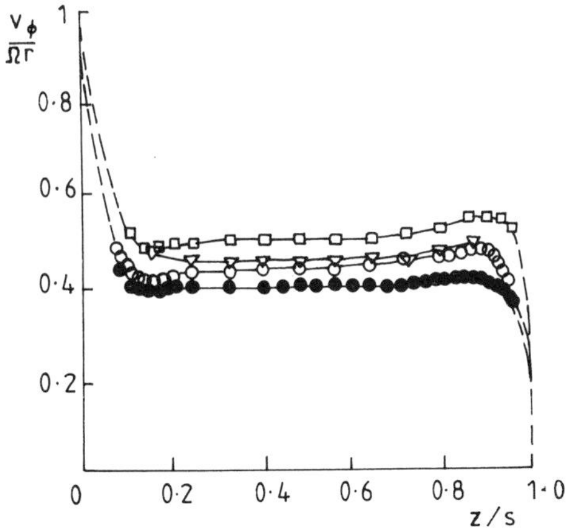

Figure 6.14 Effect of roughness on turbulent velocity profiles: $G = 0.078$, $Re_\phi = 3.1\text{x}10^6$, $x = 0.79$ (Kurokawa *et al* 1978)

Symbol:	○	●	□	▽
Rotor:	Smooth	Smooth	Rough	Rough
Stator:	Smooth	Rough	Smooth	Rough

Smooth: $k_s/b = 3.7\text{x}10^{-6}$, Rough: $k_s/b = 1.7\text{x}10^{-4}$

CHAPTER 7
Rotor-Stator Systems with Superposed Flow

When there is a superposed radial flow of fluid in the rotor-stator system, it is convenient to classify the flow into the same four regimes discussed at the beginning of Chapter 6. The boundaries of the regimes are similar to, but not identical with, those shown in Figure 6.1 (where there is no superposed flow). In particular, a superposed radial flow can cause premature transition to turbulence: indeed, at high flow rates, the flow can be turbulent when the discs are stationary. The relationship between flow rate and transition is not clear, and this subject has received little attention in the scientific literature. However, for many practical applications the flow is likely to be turbulent over most of the surface of the rotor.

Another difficult area for open systems is that of *ingress*. A superposed radial outflow implies that the *net* flow is radially outward: this does not necessarily mean that *all* the flow is outward! If the fluid "pumped" out of the system by the rotor is greater than that supplied at the centre, then ingress occurs: that is, external fluid from the surroundings is pulled into the system.

When ingress occurs from a quiescent environment, the flow structure is altered: the effect is to reduce the core rotation and to increase the moment on the rotor. For the large-clearance case, there is a tendency to cause a transition from a Batchelor-type flow to a Stewartson-type flow (see Section 6.2.1). As a consequence, the flow is more complex than for the enclosed system: the moment coefficient can be increased from a level below that of the free disc (when there is no superposed flow) to one well above. The small-clearance cases (Regimes I and III) are discussed in Sections 7.1 and 7.2, and the large-clearance ones in Sections 7.3 and 7.4. The reader is reminded at this stage that the nondimensional flow-rate, C_w, is positive for outflow and negative for inflow.

7.1 SMALL CLEARANCE: LAMINAR FLOW (Regime I)

7.1.1 Theory

To obtain a simplified picture of the flow, the approximate solution described by Soo (1958) is useful. He assumed that the radial component of velocity is the sum of a rotation-induced component (proportional to the gap Reynolds number Re_s defined in equation (6.01)) and a flow-induced one (proportional to C_w). It is convenient to write

$$\frac{v_r}{\Omega r} = -Re_s[H'(\zeta) + \phi(x)U(\zeta)], \tag{7.01}$$

$$\frac{v_\phi}{\Omega r} = V(\zeta), \tag{7.02}$$

$$\frac{v_z}{\Omega r} = 2Re_s\, G\frac{1}{x}H(\zeta), \tag{7.03}$$

where

$$\zeta = z/s\,, \qquad G = s/b \tag{7.04}$$

and

$$\phi(x) = \frac{GC_w}{2\pi Re_s^2}\frac{1}{x^2}. \tag{7.05}$$

The equation of continuity (2.35) is automatically satisfied by these expressions. They may be substituted into equations (2.36) and (2.37); the method of separation of variables can be used, in conjunction with equations (2.38), to give ordinary differential equations for $U(\zeta)$, $V(\zeta)$ and $W(\zeta)$. When $|C_w|$ is small, these equations are

$$U''' - 2Re_*^2 WU'' - 2Re_*^2 W'U' = 0, \tag{7.06}$$

$$V'' + 2Re_*^2 W'V - 2Re_*^2 WV' = 0. \tag{7.07}$$

$$W^{iv} - 2Re_*^2 WW''' - 2VV' = 0, \tag{7.08}$$

There are eight boundary conditions corresponding to conditions on the discs:

$$U(0) = W(0) = W'(0) = 0, \; V(0) = 1, \tag{7.09}$$

$$U(1) = V(1) = W(1) = W'(1) = 0. \tag{7.10}$$

To solve the equations completely, an extra condition is required. This is obtained from the continuity requirement that the mass flow-rate through the system is independent of r; this gives

$$\int_0^1 U(\zeta)\, d\zeta = 1. \tag{7.11}$$

For small values of Re_*, Soo neglected the terms involving Re_*^2 in equations (7.06) to (7.08). The equations may then be solved exactly and give

$$U(\zeta) \simeq -6\,\zeta(1 - \zeta), \tag{7.12}$$

$$V(\zeta) \simeq 1 - \zeta, \tag{7.13}$$

$$W(\zeta) \simeq -\tfrac{1}{60}\,\zeta^2(1 - \zeta)^2(3 - \zeta) \tag{7.14}$$

Figure 7.1 shows how v_r depends on ζ for different values of $\phi(x)$; it may be noted that, for "small" throughflow (that is, for $-\frac{1}{60} < \phi(x) < \frac{1}{90}$), v_r changes sign between the discs and a recirculation bubble is present. By consideration of the value of $\phi(1)$ it is easy to see that the bubble is attached to the stator when there is a superposed outflow and to the rotor when there is an

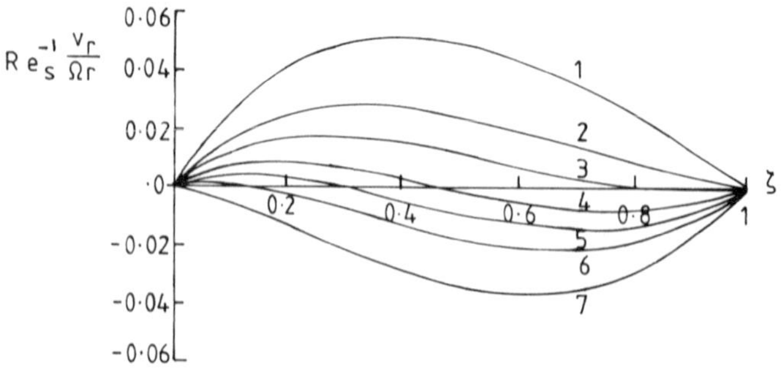

Figure 7.1 Variation of $Re_s^{-1}\frac{v_r}{\Omega r}$ with ζ: small clearances, laminar flow (Soo 1958)

Curve	1	2	3	4	5	6	7
$\Phi(x)$	$\frac{1}{45}$	$\frac{1}{90}$	$\frac{1}{180}$	0	$-\frac{1}{120}$	$-\frac{1}{60}$	$-\frac{1}{30}$

inflow; for the special case $\phi(1) = 0$, there is no radial throughflow, and a recirculation of the fluid occurs outwards along the rotor and inwards along the stator. This is summarized in Table 7.1 and typical streamline patterns are shown in Figure 7.2.

As the tangential component of velocity is sheared linearly, the shear stress is invariant with ζ and the

Table 7.1 Structure of the flow for small clearances

Type of flow	Structure of flow
"Large" outflow	outflow everywhere
"Small" outflow	outflow on rotor, separation bubble on stator (near periphery)
No net flow	no net radial outflow on rotor, radial inflow on stator
"Small" inflow	inflow on stator, separation bubble on rotor (near periphery)
"Large" inflow	inflow everywhere

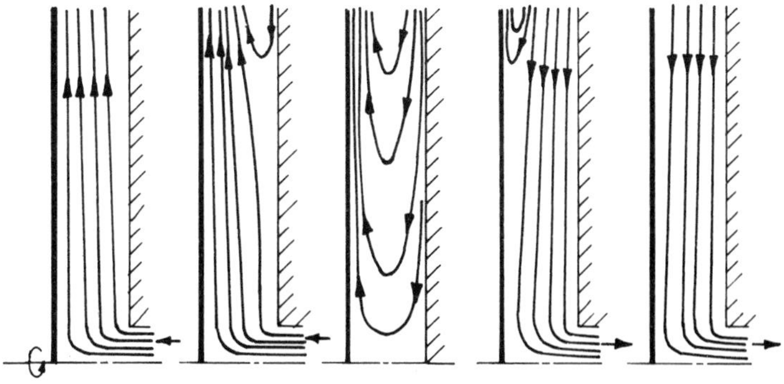

Figure 7.2 Typical streamline patterns: small clearances, laminar flow (Soo 1958)

---

moment coefficient given by equation (6.08) for Couette flow is still valid. Soo extended his solution for small values of Re_*; he found that for $Re_* < 4$, although C_M increased as Re_* increased, equation (6.08) gives values within 15% of the higher-order solution. He substituted the solution given in equations (7.12) to (7.14) into the full Navier-Stokes equations, and found that it was reasonably accurate for $|\phi(x)| < \frac{1}{60}$.

Other theoretical work has been carried out by Köhler and Müller (1971), who used a finite-difference technique to solve the boundary-layer equations, and by Szeri and Adams (1978), who solved the boundary-layer equations using a "thin-film" approximation to reduce the problem to the solution of ordinary differential equations. The latter authors were interested in applications of their work to the design of hydrostatic thrust bearings.

7.1.2 Experiment

In addition to their theoretical work, Köhler and Müller made velocity measurements, using a hot-wire anemometer, in an open rotor-stator system with discs of 400 mm diameter and a variable axial spacing. A radial outflow of air was introduced to the system through a hole of 80 mm diameter in the rotating disc. Whilst it is difficult to interpret the authors' results, it appears that most experiments were conducted for $G \simeq 0.01$ and $Re_\phi \lesssim 1200$ and that agreement between the authors' theory and their measurements was reasonable.

The calculated pressure distributions of Szeri and Adams agreed well with measurements made by other authors working in their field. The velocity distributions obtained from the thin-film approximation were also in good agreement with the LDA measurements of Szeri, Schneider, Labbe and Kaufman (1983).

7.2 SMALL CLEARANCE: TURBULENT FLOW (Regime III)

Dorfman (1961) used the integral equation (2.93) to obtain an approximate solution for the case of turbulent flow when there is no inviscid core. He assumed that

$$\int_0^s v_r v_\phi \, dz = v_{\phi,c} \frac{\dot{m}}{2\pi\rho r}, \tag{7.15}$$

where $v_{\phi,c} = v_{\phi,c}(r)$ is the value of v_ϕ in the mid-axial plane, $z = \frac{1}{2}s$. He also defined the *friction velocities* v_0^*, v_s^* by the equations

$$\tau_{\phi,0} = -\rho v_0^{*2}, \quad \tau_{\phi,s} = -\rho v_s^{*2}; \tag{7.16}$$

the minus signs occur in these expressions to ensure that $\tau_{\phi,0}$ and $\tau_{\phi,s}$ are negative. With these two assumptions, equation (2.94) reduces to

$$\dot{m} \frac{d}{dr}(r v_{\phi,c}) = 2\pi\rho r^2 (v_0^{*2} - v_s^{*2}). \tag{7.17}$$

He used $\frac{1}{7}$-power-law profiles

$$\frac{\Omega r - v_{\phi,c}}{v_0^*} = 8.74\left(\frac{s v_0^*}{2\nu}\right)^{\frac{1}{7}}, \quad \frac{v_{\phi,c}}{v_s^*} = 8.74\left(\frac{s v_s^*}{2\nu}\right)^{\frac{1}{7}}; \tag{7.18}$$

these profiles are analogous to those used in Section 4.2.2. It is convenient to define the ratio

$$\beta = \frac{v_{\phi,c}}{\Omega r} \tag{7.19}$$

and to use the nondimensional radius $x = r/b$. Equation (7.17) then becomes

$$\frac{d}{dx}(x^2\beta) = 0.168\,G^{-\frac{1}{4}}\lambda_0^{-1}\,x^{\frac{15}{4}}\left[(1-\beta)^{\frac{7}{4}} - \beta^{\frac{7}{4}}\right], \tag{7.20}$$

where

$$\lambda_0 = C_w Re_\phi^{-\frac{3}{4}}. \tag{7.21}$$

For $C_w = 0$ ($\lambda_0 = 0$), equation (7.20) implies that $\beta = 0.5$. For $\lambda_0 \to \infty$, the equation is easily solved to give $x^2\beta = x_a^2\beta_a$ where x_a, β_a are the values of x, β respectively at $r = a$, the radius at which fluid enters the system; in particular, if $\beta_a = 0$ then $\beta = 0$ for all x. For $\beta_a = 0$, the dependence of β on x for several values of λ_0 is shown in Figure 7.3. The flow structure is similar to that of Regime I (see Table 7.1).

Using equations (7.16) and (7.18), it is easy to see that

$$\frac{\tau_{\phi,0}}{\rho\Omega^2 b^2} = -0.0268\,G^{-\frac{1}{4}}\,Re_\phi^{-\frac{1}{4}}\,x^{\frac{7}{4}}\,(1-\beta)^{\frac{7}{4}}; \tag{7.22}$$

this is analogous to equation (2.98) with δ replaced by $\frac{1}{2}s$ and $\alpha = 0$. The moment coefficient for the rotor then becomes

$$C_M = 0.336\,G^{-\frac{1}{4}}Re_\phi^{-\frac{1}{4}}\int_{x_a}^{1} x^{\frac{15}{4}}(1-\beta)^{\frac{7}{4}}dx. \tag{7.23}$$

Figure 7.4 shows the dependence of $C_M G^{\frac{1}{4}} Re_\phi^{\frac{1}{4}}$ on x_a using equation (7.23), β being calculated from equation (7.20).

The two limiting cases, $\lambda_0 = 0$ ($C_w = 0$, $\beta = 0.5$) and $\lambda_0 \to \infty$ ($C_w \to \infty$, $\beta = 0$) are of interest. In these cases equation (7.23) gives

$$C_M = 0.0210\,G^{-\frac{1}{4}}\,Re_\phi^{-\frac{1}{4}}, \quad \lambda_0 \to 0, \tag{7.24}$$

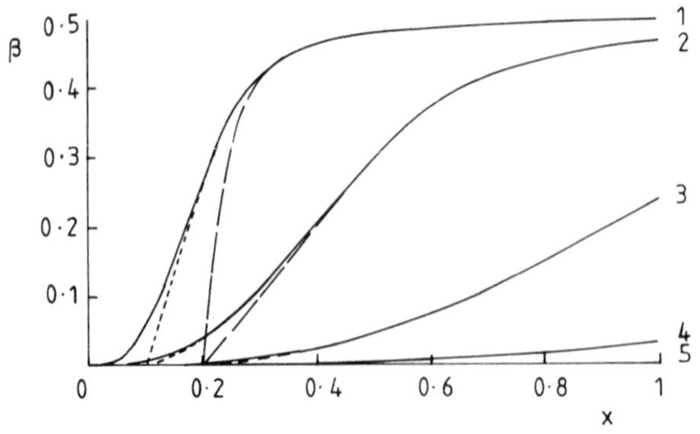

Figure 7.3 Variation of β with x: small clearances, turbulent flow, $\beta_{-} = 0$ (Dorfman 1961)

Curve	1	2	3	4	5
$G^{\frac{1}{4}}\lambda_0$	0	0.01	0.1	1	∞

——— $x_{-} = 0$, - - - - $x_{-} = 0.1$, — — — $x_{-} = 0.2$

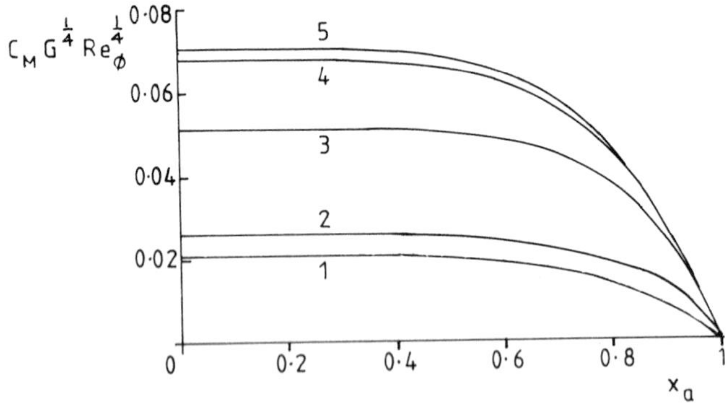

Figure 7.4 Variation of $C_M G^{\frac{1}{4}} Re_\phi^{\frac{1}{4}}$ with x_{-}: small clearances, turbulent flow, $\beta_{-} = 0$ (Dorfman 1961)

Curve	1	2	3	4	5
$G^{\frac{1}{4}}\lambda_0$	0	0.01	0.1	1	∞

$$C_M = 0.0707\, G^{-\frac{1}{4}} Re_\phi^{-\frac{1}{4}}, \qquad \lambda_0 \to \infty. \tag{7.25}$$

At the low-flow-rate limit, it is of interest to compare equation (7.24) with equation (6.07) for turbulent Couette flow. The difference between the coefficients is caused by different interpretations of the Blasius shear-stress equation for pipe flow: Soo (1958) based his result on the bulk-average velocity; Dorfman used the centre-line value. In neither case does the G-dependence agree with the experimental correlation (6.09) of Daily and Nece (1960).

The high-flow-rate limit given by equation (7.25) is created by the implicit assumption that the shear stress on the rotor cannot increase once $v_\phi = 0$ at $z = \frac{1}{2}s$: this implies that the boundary-layer thickness for v_ϕ (δ_0, say) can never be less than $\frac{1}{2}s$. In practice δ_0 can be less than this, and when this happens C_M exceeds the value given in equation (7.25). The result must, therefore, be treated with caution.

7.3 LARGE CLEARANCE: FLOW STRUCTURE (Regimes II and IV)

When the clearance is large, it is to be expected that (as for the case of no superposed flow) there is an inviscid core and, throughout most of the wheel-space, the radial mass flow occurs only in the boundary layers on the rotor and the stator. (This is similar to the flow in a rotating cavity discussed by Owen, Pincombe and Rogers 1985.)

Assuming that there is a uniform core rotation, it has been established in Chapters 3 and 4 that there can never be outflow near the stator: it is to be expected that this is also true when the core rotation depends on radius. The flow near the rotor is radially outwards or inwards when the core rotates slower or faster, respectively, than the rotor itself. For each value of x, the net mass flow-rate must equal the mass flow-rate, $\dot{m}$, of the superposed flow: equation (6.14) is therefore replaced by

$$\dot{m}_0 + \dot{m}_* = \dot{m}. \tag{7.26}$$

How the mass flow is distributed between the two discs depends on the value of $\dot{m}$ or, more specifically, on the *flow parameter*, λ. The definitions of λ for laminar and turbulent flow are

$$\lambda_{lam} = C_w Re_\phi^{-\frac{1}{2}}, \quad \lambda_{turb} = C_w Re_\phi^{-\frac{4}{5}}, \tag{7.27}$$

where a $\frac{1}{7}$-power-law profile is assumed for the turbulent case.

7.3.1 Laminar flow

In principle, it is possible to obtain a solution of the equations used in Sections 3.2.1 and 3.2.2, treating $\beta = \omega/\Omega$ as a function of x. This would involve a complicated iterative technique since the dependence of β on x is not known *a priori*, and the authors know of no serious attempt to carry out this method. Almost certainly the computing time involved for the iterations would be comparable with that required for the modern techniques for solving the full elliptic equations.

A solution can easily be found, however, employing the Ekman-layer approximation and this, as suggested by Owen (1988), should give a qualitative picture of the flow. Using a method similar to that described in Section 6.2.2, it is easy to see that equation (7.26) gives

$$1 - \beta - \beta^{\frac{1}{2}} = \frac{\lambda_{lam}}{\pi x^2}, \tag{7.28}$$

which reduces to equation (6.15) when $\lambda_{lam} = 0$. (It should be noted that, for the Ekman-layer approximation, the equations are valid locally: hence there is no loss of generality in assuming that β is a function of x.) From equation (7.28), it can be seen that $\beta = 0$ when $x = x_0$ and $\beta = 1$ when $x = x_1$ where

$$x_0 = \left(\frac{\lambda_{t\,am}}{\pi}\right)^{\frac{1}{2}}, \quad x_1 = \left(-\frac{\lambda_{t\,am}}{\pi}\right)^{\frac{1}{2}}. \qquad (7.29)$$

It follows that β can vanish only for radial outflow ($\lambda_{t\,am} > 0$) and that the tangential component of velocity in the core can exceed that on the rotor only for radial inflow ($\lambda_{t\,am} < 0$).

Equation (7.28) is quadratic in $\beta^{\frac{1}{2}}$ and, as long as $x^2 \geq 0.8\lambda_{t\,am}/\pi$,

$$\beta = \frac{1}{4}\left[-1 + \left(5 - \frac{4x_0^2}{x^2}\right)^{\frac{1}{2}}\right]^2 = \frac{1}{4}\left[-1 + \left(5 + \frac{4x_1^2}{x^2}\right)^{\frac{1}{2}}\right]^2. \qquad (7.30)$$

The positive sign for the square root is chosen to ensure that $\beta = 0.382$ (the value of β^* given by equation (6.15)) when $\lambda_{t\,am} = 0$. Clearly, if $\lambda_{t\,am} > 1.25\pi$, this equation does not give a real value for β when $x \leq 1$: in this case, the solution must be inappropriate everywhere.

For radial outflow, the mass flow-rate when $x = x_0$ is the Ekman-layer approximation to the free-disc value. A solution to equation (7.28) exists for $\sqrt{(4/5)}x_0 < x < x_0$; however, it seems plausible to assume that $\beta = 0$ for all $x < x_0$ and that the flow on the rotor corresponds to that on a free disc.

Figure 7.5 shows the variation of β/β^* (where $\beta^* = 0.382$ is the value of β given by equation (7.28) when $\lambda_{t\,am} = 0$) with $\lambda_{t\,am}/\pi x^2$. Computations have been carried out by Vaughan (1986), using the method described in Section 6.3.1, and his results for a superposed outflow are also shown in Figure 7.5. They demonstrate that the theory presented above tends to underestimate the core rotation.

Simplified versions of the streamline pattern for large ($x_0 > 1$) and small ($x_0 < 1$) outflow are shown in Figures 7.6(a) and 7.6(b) respectively. It can be seen that, for $x < x_0$, there is Stewartson-type flow with outflow on the rotor. When $x_0 < 1$ and $x > x_0$, there is a Batchelor-type flow with inflow on the stator and outflow on the rotor.

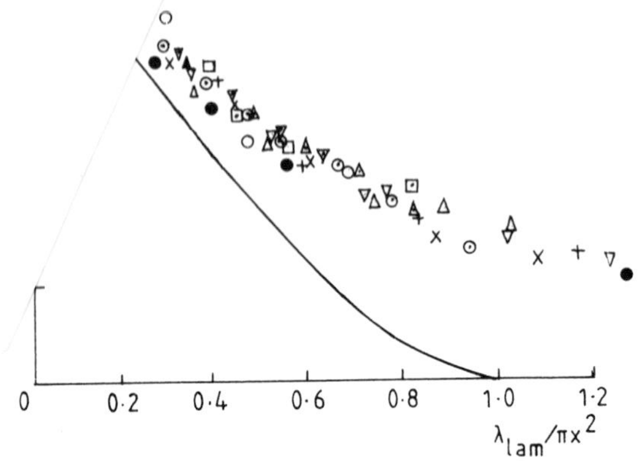

Figure 7.5 Variation of β/β^* with $\lambda_{lam}/\pi x^2$: laminar outflow, $x = 0.5$

Theory:

——— Ekman-layer approximation, equation (7.30)

• numerical predictions (Vaughan 1986)

Experiments of El-Oun and Pincombe (see Vaughan 1986):

Symbol	⊡	○	△	+	▽	X	⊙	⍋	⍒	□
$Re_\phi/10^5$	0.50	0.51	0.87	1.34	1.75	2.40	3.07	3.92	4.62	5.89

Implicit in the above theory for radial outflow is the assumption that $\beta = 0$ at the inlet. This is not always the case as the incoming fluid frequently has some swirl. Under these conditions, there is a source region in which there is a tendency for angular momentum to be conserved: the rotation in the core decays rapidly with increasing radius.

A more serious problem occurs at $x = 1$, where the fluid leaves the system. It is clear from Figures 7.6(b) and 7.6(c) that external fluid will be pulled into the wheel-space to supply the inflow on the stator: as stated

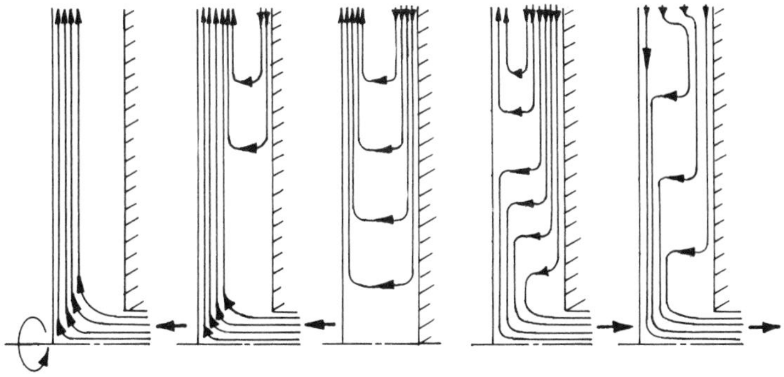

Figure 7.6 Schematic diagram of flow structures for large clearances

(a) large outflow: $x_0 > 1$, (b) small outflow: $x_0 < 1$,
(c) no superposed flow: $x_0 = x_1 = 0$
(d) small inflow: $x_1 < 1$, (e) large inflow: $x_1 > 1$

---

above, this phenomenon is referred to as ingress. The ingress can be reduced, or even prevented, by fitting a shroud on the stator: in this case, the inflow on the stator comes from a recirculating flow in the outer part of the system. The problem of ingress is discussed in detail in Chapter 9.

For small radial inflow ($\lambda_{t,m} < 0$), $x_1 < 1$ and equation (7.30) implies that there is outflow on the rotor for $x > x_1$ and inflow when $x < x_1$; for large inflow, $x_1 > 1$ and there is inflow everywhere on the rotor. (For both cases there is inflow on the stator.) The two cases are illustrated schematically in Figures 7.6(d) and 7.6(e) respectively. Difficulties arise in understanding the mechanism by which the core rotation predicted by equation (7.30) is achieved, particularly when the incoming fluid has an imposed swirl. This problem is discussed in detail for turbulent flow in Section 7.3.2.

7.3.2 Turbulent flow

The general structure of the flow is similar to that of the laminar case, but the detailed relationships are different. Explicit expressions cannot be found for the values of x_0, x_1 and β unless the approximations given in Sections 4.3.1 and 4.3.2 are used.

If the method described in Section 6.2.3 is used, equation (7.26) gives

$$\frac{49\pi}{60}\left[\alpha_0\gamma_0\,\mathrm{sgn}(1-\beta)\,|1-\beta|^{\frac{8}{5}} - \alpha_*\gamma_*\beta^{\frac{4}{5}}\right] = \frac{\lambda_{turb}}{x^{\frac{13}{5}}}, \tag{7.31}$$

where λ_{turb} is the turbulent flow parameter defined in equation (7.27). When the "improved approximation" suggested by Owen (1988) (see Section 6.2.3) is made, $\alpha_0 = \alpha_0(\beta)$, $\gamma_0 = \gamma_0(\beta)$, $\alpha_* = \alpha_0(1)$ and $\gamma_* = \gamma_0(1)$; $\alpha_0(\beta)$ and $\gamma_0(\beta)$ are given by equations (4.67) and (4.68) respectively. Equation (7.31) reduces to equation (6.18) when $\lambda_{turb} = 0$.

If the Ekman-layer approximation is made for both discs (and $\alpha_0 = \alpha_* = \alpha_0(1) = 0.553$ and $\gamma_0 = \gamma_* = \gamma_0(1) = 0.0983$), equation (7.31) simplifies to

$$\mathrm{sgn}(1-\beta)\,|1-\beta|^{\frac{8}{5}} - \beta^{\frac{4}{5}} = \frac{\lambda_{turb}}{0.1395x^{\frac{13}{5}}}, \tag{7.32}$$

For $\lambda_{turb} = 0$, this equation reduces to equation (6.19) and $\beta = \beta^* = 0.382$. The variation of β/β^* with $\lambda_{turb}x^{-\frac{13}{5}}$, according to equation (7.32), is shown in Figure 7.7 for radial outflow ($\lambda_{turb} > 0$).

As discussed in Section 7.3.1, $x = x_0$ when $\beta = 0$, and $x = x_1$ when $\beta = 1$. It follows from equation (7.31) (using equations (4.67) and (4.68)) that

$$x_0 = \left(\frac{\lambda_{turb}}{0.219}\right)^{\frac{5}{13}}, \qquad x_1 = \left(-\frac{\lambda_{turb}}{0.140}\right)^{\frac{5}{13}}. \tag{7.33}$$

As in the laminar case, β can vanish only for outflow; then the mass flow-rate on the rotor at $x = x_0$ is equal to

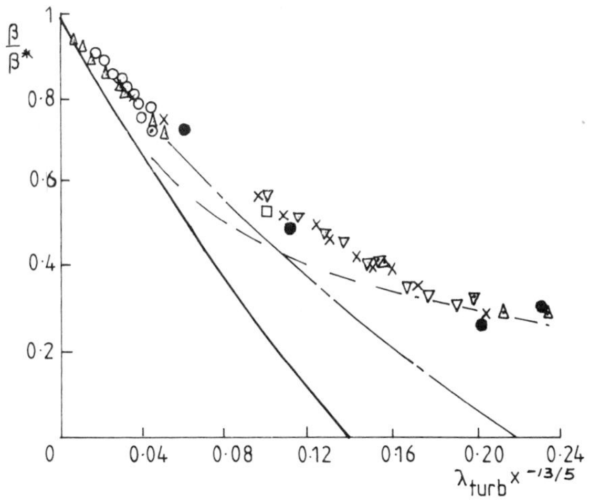

Figure 7.7 Variation of β/β^* with $\lambda_{turb}/x^{\frac{13}{5}}$: turbulent outflow

— · — · —	"improved approximation", equation (7.31)
————	Ekman-layer approximation, equation (7.32)
— — —	empirical correlation (7.37) (Daily *et al* 1964)
●	numerical predictions (Vaughan 1986)

Experiments of El-Oun and Pincombe (see Vaughan 1986):

Symbol	⊙	◬	△	×	⩢	□	▽	○
$Re_\phi/10^6$	0.69	0.76	0.84	0.93	1.07	1.10	1.12	1.122

that entrained by the free disc. Similarly, β can be unity only for inflow. The variation of β/β^* (where $\beta^* = 0.43$ is the solution of equation (7.32) when $\lambda_{turb} = 0$) with $\lambda_{turb} x^{-\frac{13}{5}}$ is shown in Figure 7.7 for radial outflow, and that of β with $x/|\lambda_{turb}|^{\frac{5}{13}}$ in Figure 7.8 for both outflow and inflow. It should be noted that the method fails if $\beta \gtrsim 4.37$, as can be seen from equation (4.67) for α.

Although equation (7.31) is an implicit equation for β in terms of $\lambda_{turb} x^{-\frac{13}{5}}$, it may be regarded, for graph-drawing purposes, as an explicit equation for

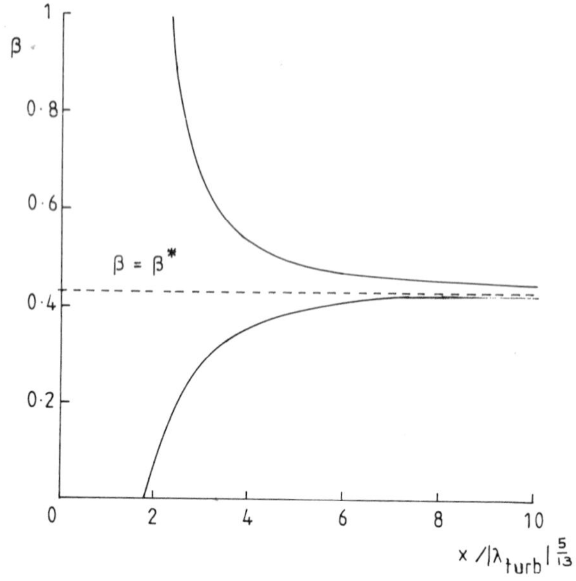

Figure 7.8 Variation of β with $x/|\lambda_{turb}|^{\frac{5}{13}}$

———— "improved approximation", equation (7.32)

$\lambda_{turb}x^{-\frac{13}{5}}$ in terms of β. Explicit expressions for β can be found using the approximations for $\epsilon_{m,0}$ given in equations (4.74) and (4.75). Then, for outflow,

$$\frac{\beta}{\beta^*} \simeq \left[3.35 - 2.35\left(1 + 0.99\left(\frac{x_0}{x}\right)^{\frac{13}{5}}\right)^{\frac{1}{2}}\right]^{\frac{5}{4}}, \quad x \geq x_0 \qquad (7.34)$$

and, for inflow,

$$\frac{\beta}{\beta^*} \simeq \left[1 + 1.6\left(1.31\left(\frac{x_1}{x}\right)^{\frac{13}{5}} - 1\right)^{\frac{1}{2}}\right]^{\frac{5}{4}}, \quad 0.49x_1 \leq x \leq x_1 \qquad (7.35)$$

$$\frac{\beta}{\beta^*} \simeq \left[3.35 - 2.35\left(1 - 0.63\left(\frac{x_1}{x}\right)^{\frac{13}{5}}\right)^{\frac{1}{2}}\right]^{\frac{5}{4}}, \quad x \geq x_1 . \qquad (7.36)$$

(The lower limit on x for inflow occurs because the approximation given in equation (4.75) is inadequate for $\beta > 3.5$.) Equations (7.34) and (7.36) are both valid for

$\beta < 1$ and are exactly equivalent; the two forms are given for convenience.

It is legitimate, when using the Ekman-layer approximation for both discs, to match the local mass flow-rates on the rotor and the stator. However, there is some doubt about its justification when the full equations are used for the rotor: the solutions were obtained assuming that the core rotation is independent of x and, for the case discussed here, this assumption is not true. The justification ultimately depends on how well the solution agrees with experimental data, and a discussion of this is deferred to the next section.

Numerical predictions by Vaughan (1986) for radial outflow, using the method described in Section 6.4.1, are also shown in Figure 7.7. It can be seen that the "improved approximation" discussed above is in good agreement with the computed values when $\lambda_{turb} x^{-\frac{13}{5}} < 0.1$.

The general structure of the flow is similar to that for the laminar case, shown schematically in Figure 7.6. Recently, Dr Vaughan has achieved considerable reduction in computing time by the inclusion of a multigrid algorithm (see Lonsdale 1988) in his turbulent elliptic solver. This program was used by Drs Ong and Vaughan (private communication) to predict the flow for a number of inflow cases in which $G = 0.1$, $Re_\phi = 10^6$ and $\lambda_{turb} = -0.08$. Some of their results are shown in Figures 7.9 and 7.10: the predicted streamlines are shown in Figure 7.9, and the corresponding variation of β with x in Figure 7.10. The computations were carried out for a uniform source flow with three different values of the inlet *swirl fraction*, c, of the fluid (where c is the value of $v_{\phi,c}/\Omega r$ at $x = 1$). Computations were conducted for the cases where there was no peripheral shroud and also where there was a shroud attached either to the rotor or to the stator. For the cases with a shroud, the clearance

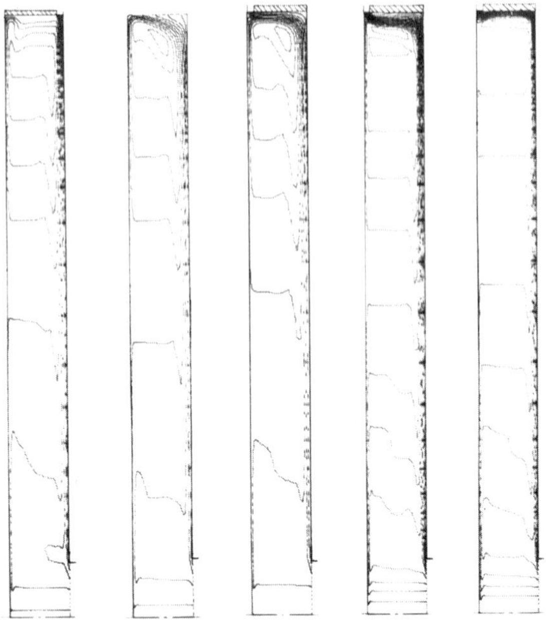

c	0.27	0.27	0.27	0.64	1.5
shroud	R	N	S	S	S

(R: rotating shroud, N: no shroud, S: stationary shroud)

Figure 7.9 Computed streamlines in a rotor-stator system: turbulent inflow, $G = 0.1$, $G_c = 0.01$, $Re_\varphi = 10^6$, $\lambda_{turb} = -0.08$

---

ratio was $G_c = s_c/b = 0.01$ (s_c being the axial clearance between the shroud and the stator or the rotor).

It can be seen from Figure 7.9 that all the incoming fluid (or, for the case $c = 1.5$, most of it) is drawn very quickly into the boundary layer on the stator. This fluid subsequently leaves the stator, crosses the core (with little or no radial velocity) and is entrained into the boundary layer on the rotor. On the rotor, the flow is radially outward where $x > x_1$ (for which $\beta < 0$) and inward where $x < x_1$ ($\beta > 0$).

The position at which $\beta = 1$ does not depend on the shroud

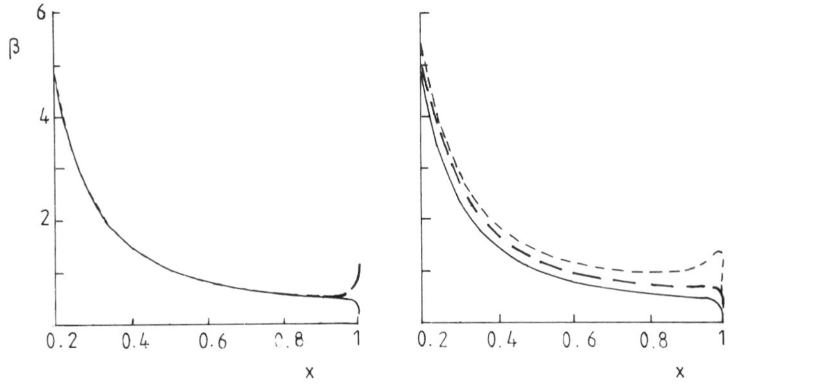

(a) Effect of shroud $c = 0.27$ (b) Effect of swirl fraction Stationary shroud

Figure 7.10 Computed radial variation of β in a rotor-stator system: turbulent inflow, $G = 0.1$, $G_c = 0.01$, $Re_\phi = 10^6$, $\lambda_{turb} = -0.08$

	(a)	(b)
————————	Stationary shroud	$c = 0.27$
— — —	No shroud	$c = 0.64$
- - - - -	Rotating shroud	$c = 1.5$

(see Figure 7.10 (a)), but it is strongly dependent on the value of c (see Figure 7.10 (b)). It should be noted that not only does the position of x_1 affect the direction of flow on the rotor, it also affects the radial pressure distribution inside the system and the frictional moment exerted on the rotor. (It may be of interest to note that equation (7.33) gives $x_1 = 0.806$ for $\lambda_{turb} = -0.08$. However, the velocity distribution predicted by equation (7.31) is significantly too high for $x < 0.8$.)

7.3.3 Experiment

Daily, Ernst and Asbedian (1964), using equipment essentially the same as that employed by Daily and Nece (1960) (see Chapter 6), measured the tangential component of velocity in the core for a superposed outflow of air. They correlated their results by the relation

$$\frac{\beta}{\beta^*} = \left(1 + 12.74 \frac{\lambda_{turb}}{x^{13/5}}\right)^{-1} , \qquad (7.37)$$

which is shown in Figure 7.7.

The experimental data shown in Figures 7.5 and 7.7 were obtained by El-Oun and Pincombe from LDA measurements in the rotor-stator rig described in Section 8.3.2: further details are given by Vaughan (1986). Their results are in good agreement with Vaughan's numerical predictions for both laminar and turbulent flow. It is clear that in the neighbourhood of $\lambda_{turb} x^{-\frac{13}{5}} = 0.22$ there is a small core rotation in practice ($\beta \simeq 0.1$), although the "improved approximation" (discussed in Section 7.3.2) predicts that $\beta = 0$. This means that a mass flow-rate corresponding to $C_w = C_{w,rd}$ is insufficient to suppress completely the core rotation.

It is shown below that the approximate theory, though inadequate for predicting the core velocity near the axis of rotation, is satisfactory for predicting the moment coefficient (and, by analogy, the Nusselt number).

7.4 LARGE CLEARANCE: MOMENT COEFFICIENTS FOR TURBULENT FLOW (Regime IV)

7.4.1 Approximate solution for a small radial outflow

It is assumed that the radial outflow is small enough to ensure that $x_0 < 1$, x_0 being the value of x at which $\beta = 0$ (as given by equation (7.33)). It is also assumed that it is large enough to ensure that ingress does not occur.

The moment coefficient, C_M, on the rotor can be expressed as

$$C_M = C_{M,1} + C_{M,2} , \qquad (7.38)$$

where

$$C_{M,1} = -4\pi \int_0^{x_0} x^2 \frac{\tau_{\phi,0}}{\rho \Omega^2 b^2} dx , \qquad (7.39)$$

$$C_{M,2} = -4\pi \int_{x_0}^{1} x^2 \frac{\tau_{\phi,0}}{\rho\Omega^2 b^2} dx . \tag{7.40}$$

For $x < x_0$, where $\beta = 0$, equation (4.18) is appropriate and

$$C_{M,1} Re_\phi^{\frac{1}{5}} = 0.0729 \, x_0^{\frac{23}{5}} . \tag{7.41}$$

For $x > x_0$, equations (2.74), (4.69) and (4.73) give

$$\frac{\tau_{\phi,0}}{\rho\Omega^2 b^2} = -\frac{23 + 37\beta}{60\pi} \epsilon_{m,0} \, x^{\frac{8}{5}} Re_\phi^{-\frac{1}{5}}, \qquad x > x_0 , \tag{7.42}$$

where $\epsilon_{m,0}$ is given by equation (4.70). Using the approximation given in equation (4.74) for $\epsilon_{m,0}$, equation (7.40) becomes

$$C_{M,2} Re_\phi^{\frac{1}{5}} = \int_{x_0}^{1} (23 + 37\beta)(0.01027\beta^{\frac{8}{5}} - 0.02573\beta^{\frac{4}{5}} + 0.0152) \, x^{\frac{18}{5}} dx , \tag{7.43}$$

where β can be approximated by equation (7.34) and the integral can be evaluated numerically. For the particular case in which $C_w = 0$ (so that $x_0 = 0$ and hence $C_{M,1} = 0$), $C_M^* Re_\phi^{\frac{1}{5}} = 0.040$, which is approximately 55% of the free-disc value. This should be compared with the values of about 50% for $C_{M,min}^* / C_{M,fd}$ found for an enclosed system (see Section 6.4.2).

The variation of $\tau_{\phi,0}/\rho\Omega^2 b^2$ with x, using equations (4.18) and (7.42), is shown in Figure 7.11. Also shown are the computations of Vaughan (1986) for a shrouded rotor-stator system with $G = 0.069$, $C_w = 1000$ and $Re_\phi = 8 \times 10^5$. For $x \leq x_0 = 0.389$, the numerical curve is higher than that given by equation (4.18); for $x \geq x_0$, the numerical curve is slightly lower than that predicted by equation (7.42). (The reduction in $\tau_{\phi,0}$ shown by the numerical solution near $x = 1$ is attributed to the reduction in the flow in the boundary layer on the rotor as fluid is recirculated to the stator via the boundary

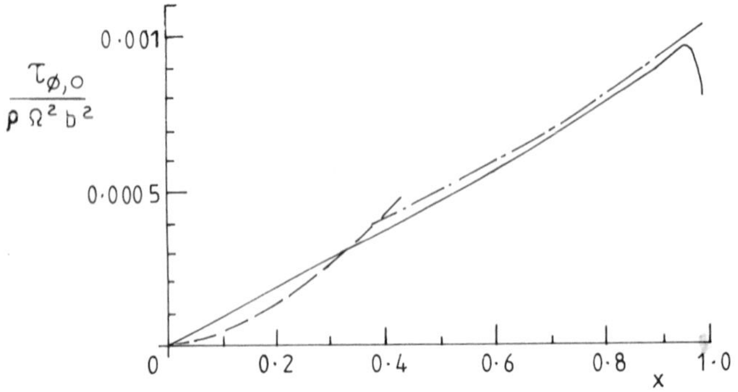

Figure 7.11 Variation of $\tau_{\phi,0}/\rho\Omega^2 b^2$ with x for a small turbulent outflow: $G = 0.069$, $C_w = 1000$, $Re_\phi = 8\times10^5$

— — — the free disc, equation (4.18)
— · — · — "improved approximation", equation (7.42)
———— numerical predictions (Vaughan 1986)

---

layer on the shroud.)

The measured moment coefficients of Daily *et al* (1964), for $G = 0.0276$, 0.069 and 0.124, $2\times10^6 \le Re_\phi < 10^7$ and $0 \le \lambda_{turb} < 0.06$ were correlated by

$$C_M = C_M^* \left(1 + 13.9\,\beta^* \lambda_{turb} G^{-\frac{1}{8}}\right). \qquad (7.44)$$

The authors' values of C_M^* and β^* are given in Table 6.1.

Figure 7.12 shows the variation of $C_M^* Re_\phi^{\frac{1}{5}}$ with λ_{turb}, using the approximate theory described above, and the computations of Vaughan (1986) for $G = 0.069$; these are within 5% of each other. Also shown are the experimental correlations of Daily *et al* given by equation (7.44), and these are in reasonable agreement with the theoretical results. Subsequent computations by Chew and Vaughan (1988) show that $C_M Re_\phi^{-\frac{1}{5}}$ does indeed become asymptotic to the free-disc value as $\lambda_{turb} \to 0.22$.

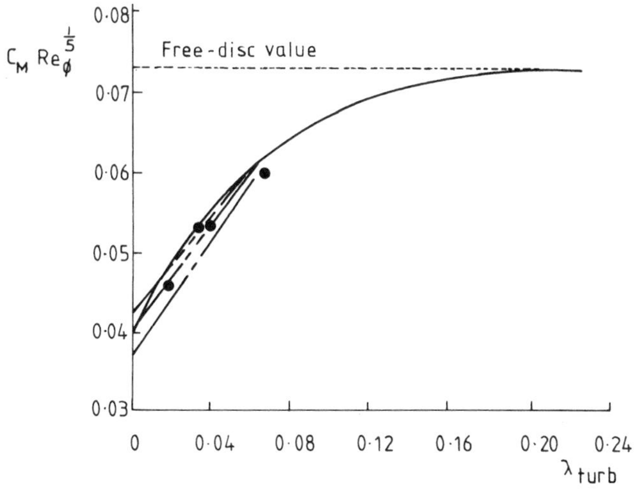

Figure 7.12 Variation of $C_M Re_\phi^{\frac{1}{5}}$ with λ_{turb} for a small outflow

———————	"improved approximation", equation (7.37)
— · — · — ·	$G = 0.0276$ } empirical correlation (7.44) (Daily *et al* 1964)
— · · — · · —	$G = 0.069$ }
— · · · — · · ·	$G = 0.124$ }
•	numerical predictions: $G = 0.069$ (Vaughan 1986)

7.4.2 Radial outflow in unshrouded systems

An unshrouded rotor-stator system was studied theoretically and experimentally by Bayley and Owen (1969) and by Owen, Haynes and Bayley (1974). In the experiments, the rotor was 762 mm diameter and could be rotated up to 4000 rev/min. Air was admitted through a hole of 102 mm diameter at the centre of the stator, and the gap between the rotor and the stator could be varied from 2 to 230 mm. Aerodynamic probes were used to measure velocities at the periphery, the radial distribution of pressure was measured by taps in the stator, and the moment on the rotor was determined using a built-in torquemeter. In the theoretical study, the turbulent boundary-layer equations were solved numerically using the finite-difference method of Spalding and Patankar (1967), and a mixing-length model was used to evaluate the shear stresses.

Figure 7.13 shows the variation of C_M with G for $Re_\phi = 9 \times 10^5$ and $0.53 \le C_w/10^4 \le 7.4$. For this value of Re_ϕ, equation (4.17) gives $C_{w,fd} = 1.3 \times 10^4$; Figure 7.13 shows that for $C_w < C_{w,fd}$ the moment coefficient can be less than the free-disc value. For an unshrouded system, ingress can occur and fluid can be readily entrained from the surroundings. Consequently the core rotation is attenuated and, at the larger gap ratios, it can be suppressed completely and Stewartson-type flow occurs. Under these

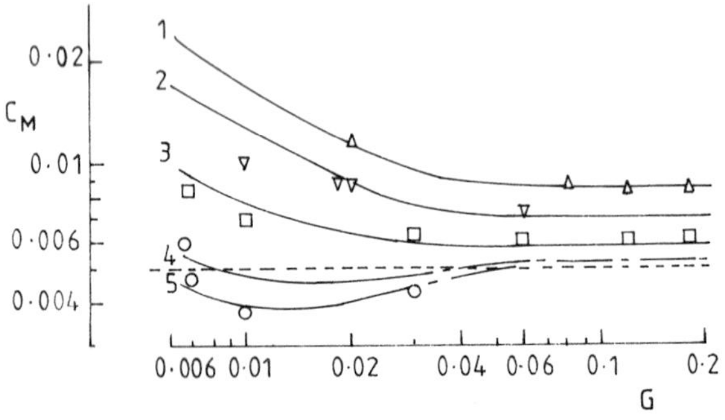

Figure 7.13 Variation of C_M with G for radial inflow in an unshrouded system: $Re_\phi = 9 \times 10^5$ (Owen *et al* 1974)

Theory: ———— Numerical solution

- - - - free-disc correlation (7.47)

Experiment:	△	▽	□		○
Curve	1	2	3	4	5
C_w	7.4×10^4	4.7×10^4	2.2×10^4	1.3×10^4	0.53×10^4

conditions, at larger gap ratios, $C_M \to C_{M,fd}$ even when $C_w < C_{w,fd}$. For $C_w > C_{w,fd}$, C_M increases as G decreases and $C_M > C_{M,fd}$ for all gap ratios.

The authors proposed that, for an unshrouded system, the limiting gap ratio, G_{lim}, beyond which the moment on the rotor would be unaffected by the presence of the stator, could be estimated by

$$G_{lim} = 1.05\, Re_\phi^{0.2}. \tag{7.45}$$

For $Re_\phi = 9\text{x}10^5$, this gives $G_{lim} = 0.068$ which is consistent with the results shown in Figure 7.11. The numerical procedures were invalid for the case where recirculation occurred, and no solutions were obtained for $G > 0.055$. For comparison with experimental data obtained at the larger gap ratios, computations were carried out by setting $G = 0.055$; for $G < 0.055$, the actual values of G were used.

For the case where $C_M \geq C_{M,fd}$ (that is, for $C_w \geq C_{w,fd}$ or $G \geq G_{lim}$), Owen and Haynes (1976) obtained correlations of the measured moment coefficients using two asymptotic results. For $C_w/G\, Re_\phi > 10$, an approximate solution of the momentum-integral equations gave

$$C_M = 0.0553 \left(\frac{C_w}{G}\right)^{\frac{4}{5}} Re_\phi^{-1}; \tag{7.46}$$

the limit for large Re_ϕ was based on their measured free-disc value (see Section 4.2.5)

$$C_{M,fd} = 0.0655\, Re_\phi^{-0.186}. \tag{7.47}$$

Using Truckenbrodt's (1954) solution for impinging flow on a rotating disc (see Section 4.3.3), the following correlation was proposed:

$$C_M = (C_{M,1}^6 + C_{M,2}^6)^{\frac{1}{6}} \tag{7.48}$$

where

$$C_{M,1} = 0.0553 \left(\frac{C_w}{G}\right)^{0.8} Re_\phi^{-1} \tag{7.49}$$

and

$$C_{M,2} = 0.0655Re_{\phi}^{-0.186}\left[1 + 12.4\frac{C_w}{Re_{\phi}}\right]. \qquad (7.50)$$

The correlation was tested for $b/a = 7.5$, $0.01 \leq G \leq 0.18$, $1.4 \leq C_w/10^4 \leq 9.8$ and $0 < Re_{\phi} \leq 4\text{x}10^6$, and the variation of C_M with Re_{ϕ} for $G = 0.02$ and $G = 0.18$ is shown in Figure 7.14. The agreement between the correlation, the measured values and the numerical solutions are better at the smaller gap ratio.

Like all correlations, caution must be observed in extrapolating outside the range of the tests. In particular, equation (7.48) implies that $C_M \geq C_{M,fd}$ whereas, as noted above, C_M can be less than $C_{M,fd}$ even for an unshrouded system. The effect of fitting a peripheral shroud is discussed below.

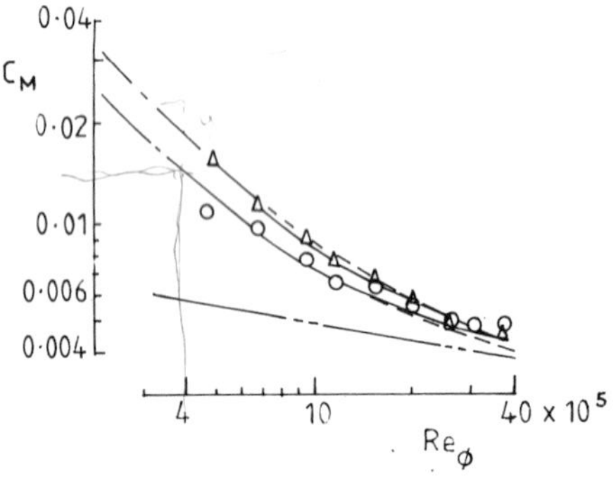

(a) $G = 0.02$

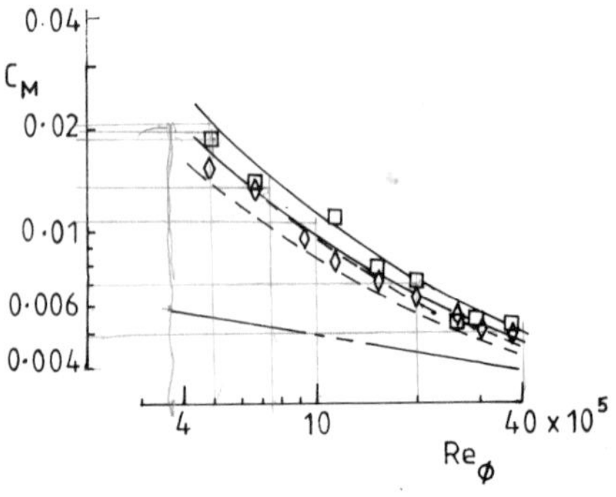

(b) $G = 0.18$

Figure 7.14 Variation of C_M with Re_{ϕ} for a turbulent unshrouded system (Owen and Haynes 1976)

Theory:	————————	empirical correlation (7.48)
	— — — —	numerical solution
	— · — · —	approximation solution (7.46)
	— ·· — ··	free-disc correlation (7.47)

Experiment:	Symbol	○	△	◊	□
	C_w	$3.2\text{x}10^4$	$4.7\text{x}10^4$	$7.4\text{x}10^4$	$9.7\text{x}10^4$

7.4.3 Radial outflow in shrouded systems

Experiments were conducted on a shrouded system by Bayley and Owen (1970) and by Haynes and Owen (1975). The basic apparatus was the same as that described in Section 7.4.2, and the shroud was attached to the stator leaving an adjustable axial clearance, s_c, between the shroud and the rotor (see Figure 7.15). For the experiments, the clearance was adjusted to produce variations of G_c (where $G_c = s_c/b$) in the range $0.0033 \leq G_c \leq G$ where $0.06 \leq G \leq 0.12$.

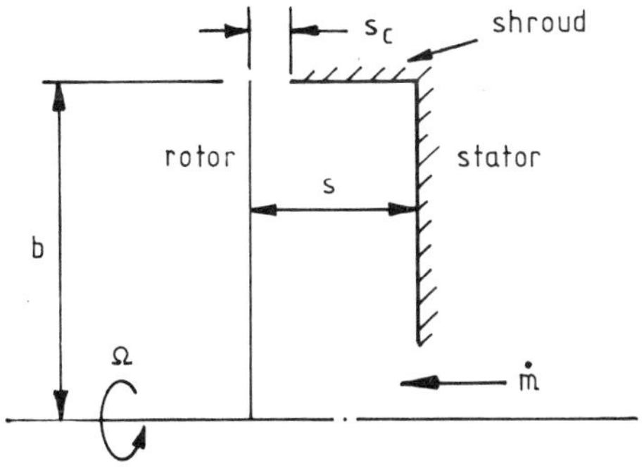

Figure 7.15
Schematic diagram of a shrouded rotor-stator system

The step-change from an axial gap of s inside the wheel-space to a clearance s_c at the shroud cannot be modelled by the boundary-layer equations. However, Haynes and Owen avoided the problem by assuming a smooth transition of the computational domain upstream of the shroud. Whilst not providing an accurate description of the flow near the shroud, the technique was effective in predicting the effect of G_c on both the pressure distribution and the frictional moment. For the reasons discussed in Section 7.4.2, the computations were conducted at a fixed value of $G = 0.055$.

Figure 7.16 shows the variation of C_M with Re_ϕ for various values of G, G_c and C_w, and there is generally good agreement between the computations of Haynes and Owen

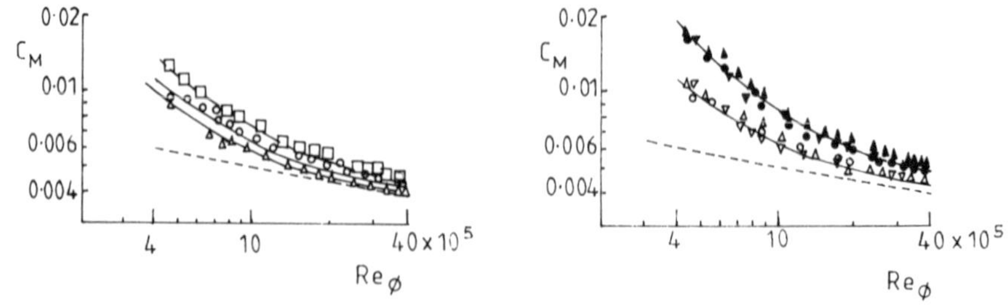

(a) (b)

Figure 7.16 Variation of C_M with Re_ϕ for a shrouded system (Haynes and Owen 1975)

———— Numerical solution

- - - - - Free disc, correlation (7.47)

Experiment:

(a) $G = 0.06$, $C_w = 2.4\text{x}10^4$

□ $G_c = 0.0033$, ○ $G_c = 0.0067$, △ No shroud

(b) $G_c = 0.0067$

△ $G = 0.18$, ▽ $G = 0.12$, ○ $G = 0.06$

Open symbols, $C_w = 2.4\text{x}10^4$; Solid symbols, $C_w = 5.3\text{x}10^4$

---

and their measured values of C_M. In particular, Figure 7.16(a), for $C_w = 2.4\text{x}10^4$ and $G = 0.06$, shows that C_M increases as G_c decreases; Figure 7.16(b), for $G_c = 0.0067$, shows that C_M increases as C_w increases and that the effect of gap ratio is weak for $0.06 < G < 0.18$. The flow rates were high enough to ensure that the moment coefficient was greater than the free-disc value.

Gosman, Lockwood and Loughhead (1976) obtained numerical solutions of the elliptic equations using the method of Gosman, Pun, Runchal, Spalding and Wolfshtein (1969), together with the $k-\epsilon$ turbulence model. In this method, the finite-difference equations for vorticity and stream-function are solved rather than the equations for the "primitive variables" (velocity and pressure). Although the authors encountered stability problems at large values of Re_ϕ, their computed moment coefficients for the shrouded rotating-disc system were in good agreement with the experimental data of Bayley and Owen (1970).

Gosman, Koosinlin, Lockwood and Spalding (1976) obtained numerical solutions of the elliptic equations for the "primitive variables". They introduced a number of modifications to improve stability and convergence at large values of Re_ϕ, and their predicted moment coefficients and velocity distributions were in good agreement with the measurements of Bayley and Owen for both shrouded and unshrouded systems.

The effects on frictional windage caused by bolt heads and other protrusions on the rotor have been measured by Dibelius, Radke and Ziemann (1984), Zimmermann, Firsching, Dibelius and Ziemann (1986), Graber, Daniels and Johnson (1987) and Millward and Robinson (1989). Whilst bolt heads can cause a large increase in C_M, the effect can be reduced significantly by enclosing the heads in an axisymmetric casing. In fact, Millward and Robinson found that, by encasing the bolts near the periphery of the rotor, the windage was *less* than that for a plane disc!

7.4.4 Radial inflow

Bayley and Conway (1964) measured velocity and pressure distributions, and the moment on the stator, in an open rotor-stator system with a radial inflow of air. The rotor and stator were 762 mm diameter, and the air was extracted through a hole of 102 mm diameter in the centre of the stator. Experiments were conducted for $0.004 \leq G \leq 0.06$, $0 \leq |C_w| \leq 10^4$ and $0 \leq Re_\phi \leq 4\text{x}10^6$.

It was observed in Section 7.3.2 that the value of the inlet swirl-fraction, c, affects the size of the region in which radial outflow occurs on the rotor, and this in turn affects the magnitude of the moment coefficient. Dibelius *et al* (1984) and Graber *et al* (1987), whose measurements on discs with protrusions are referred to above, also conducted experiments in rotor-stator systems with a radial inflow of air. Dibelius *et al* used fixed-angle swirl vanes (for which c depended on the ratio $|C_w|/Re_\phi$) and in the experiments of Graber *et al*, $c \simeq \frac{1}{2}$. Both groups of research workers found that C_M decreased when $|C_w|$ was increased. However, when Dibelius *et al* removed the vanes (so that $c \simeq 0$), they found that C_M increased when $|C_w|$ was increased. The value of the inlet swirl fraction is clearly an important parameter for radial inflow.

CHAPTER 8
Heat Transfer in Rotor-Stator Systems

Most of the measurements of heat transfer in a rotor-stator system have been made on the rotor (which is rotating with angular speed Ω and has a temperature T_0). For large clearances (see Chapter 6) there is an inviscid core in which the tangential component of velocity $v_{\phi,c}$ and the temperature T_c are assumed to be independent of z. For small clearances, it is often convenient to take $v_{\phi,c}$ and T_c as the values of v_ϕ and T respectively in the mid-axial plane, $z = \frac{1}{2}s$.

When there is no superposed flow, most experimentalists define a Nusselt number, Nu, using a temperature difference $\Delta T = T_0 - T_c$ (in spite of the practical difficulties involved in making measurements in the middle of the fluid!). The Nusselt numbers then become

$$Nu = \frac{rq_0}{k(T_0 - T_c)}, \quad Nu_{av} = \frac{bq_{0,av}}{k(T_0 - T_c)_{av}}. \tag{8.01}$$

When there is a superposed flow, the temperature of the incoming fluid is taken to be T_I, and it is natural to take $\Delta T = T_0 - T_I$ in the definition of the Nusselt number: this gives

$$Nu = \frac{rq_0}{k(T_0 - T_I)}, \quad Nu_{av} = \frac{bq_{0,av}}{k(T_0 - T_I)_{av}}. \tag{8.02}$$

However, some authors use the "adiabatic temperature difference" $\Delta T = T_0 - T_{0,ad}$ where

$q_0 = \frac{q}{A} = h\Delta T \Rightarrow Nu = \frac{rh\Delta T}{k\Delta T} = \frac{rh}{k}$

$Nu_{av} = \frac{bh}{k}$

$$T_{0,ad} = T_I + R^r \frac{\Omega^2 r^2}{2c_p}, \tag{8.03}$$

by analogy with equation (2.121). The Nusselt numbers in these cases are written as

$$Nu^* = \frac{rq_0}{k(T_0 - T_{0,ad})}, \quad Nu^*_{av} = \frac{bq_{0,av}}{k(T_0 - T_{0,ad})_{av}}. \tag{8.04}$$

There is a basic difficulty in developing a theoretical model for the heat transfer in a rotor-stator system: whether or not there is a superposed flow through the wheel-space, it is necessary to know the temperature T_c to compute q_0 (see, for example, equation (2.122) with T_∞ replaced by T_c). Even if equation (8.04) is used for the Nusselt number and dissipation is neglected, it can be seen from equation (2.125) that the correction factor χ must be determined; this is discussed in Sections 5.2 and 5.3 where it is shown that the radial distribution of T_c (or T_∞) must be known. For a small clearance, when there is no inviscid core, it may be possible to use the theory of Chapter 5 with T_∞ replaced by T_s, the temperature of the stator; as far as the authors are aware, this has not been attempted.

In view of the difficulty in determining the core temperature, it is useful to compare experimental measurements in the rotor-stator system with the results for the free disc given in equations (5.67) and (5.77).

8.1 ROTOR-STATOR SYSTEMS WITH NO SUPERPOSED FLOW

Comparatively few experiments have been conducted for an enclosed system with no superposed flow of coolant. For such systems, frictional heating and heat transfer from the rotor is convected to the adjacent fluid which in turn transfers the heat to the stationary casing.

Nikitenko (1963) conducted experiments with an air-filled enclosed system where the horizontal rotor and the stator were both isothermal; the rotor was 600 mm diameter and instrumented with fluxmeters. Experiments were conducted for $0.018 \leq G \leq 0.085$ and $Re_\varphi \leq 10^6$, but there are no details in the paper about heat balances of how the temperatures of the fluid core were measured. Nikitenko found no effect of gap ratio and correlated his results by

$$Nu_{lam} = 0.675\,(x^2 Re_\varphi)^{\frac{1}{2}}, \quad Nu_{turb} = 0.0217\,(x^2 Re_\varphi)^{\frac{4}{5}}. \quad (8.05)$$

The laminar value is over 80% greater, and the turbulent value is about 12% greater, than the free-disc case values corresponding to equations (5.67) and (5.75) with $Pr = 0.71$ and $n = 0$. Nikitenko also obtained similar correlations for the stator with constants of 0.364 and 0.0178 in the equations equivalent to (8.05).

Shchukin and Olimpiev (1975) measured the average Nusselt numbers in an air-filled enclosed system with a vertical rotor of 930 mm diameter, which was 33 mm thick and was filled with water. The stator was heated electrically to produce temperature differences between the air and the rotor from 270 to 500°C; the heat flux through the rotor was determined by solving the conduction equation. During the tests, which were conducted with $G = 0.0646$, the radial temperature distribution on the rotor varied as r^n with $0 < n < 0.6$. Measurements were made in the transition region ($10^5 < Re_\varphi < 10^6$) and in the turbulent region ($10^6 < Re_\varphi < 3\mathrm{x}10^6$).

For a temperature profile with $n = 0.25$, their results for turbulent flow were correlated by

$$Nu^*_{av,\,turb} = 0.0168\,Re_\varphi^{\frac{4}{5}}, \quad (8.06)$$

which is approximately 11% higher than the free-disc value corresponding to equation (5.75).

8.2 ROTOR-STATOR SYSTEMS WITH SUPERPOSED FLOW

8.2.1 Radial outflow through small clearances

For laminar flow Soo, Besant and Sarafa (1962) used Soo's (1958) first-order solution of the Navier-Stokes equations (see Section 7.1.1) to solve the energy equation (2.39). They considered the case of an isothermal rotor and a stator that was either isothermal or adiabatic. However, as their solutions are valid only over a very restricted range of G, C_w and Re_ϕ, the results are of little practical value.

Kreith and Viviand (1966) obtained series solutions of the laminar boundary-layer equations for small values of both Re_* and GRe_ϕ/C_w. As the radial component of velocity is approximately parabolic under these conditions, the classical Graetz pipe-flow solutions (see Schlichting 1979) were used to determine the Nusselt numbers. The results (again of little practical value) were in good agreement with the authors' measured values.

For turbulent flow, Dorfman (1961) applied the Reynolds analogy to his solution of the momentum-integral equations (see Section 7.2). He assumed that the temperature and the tangential component of velocity have similar distributions, a result which is valid only when there is a quadratic temperature on the disc. His results were consistent with the measured Nusselt numbers of Kapinos (1957) when $G < 0.03$.

8.2.2 Unshrouded systems with radial outflow

Kreith, Doughman and Kozlowski (1963) made mass transfer measurements on a naphthalene-coated disc rotating close to a stator through the centre of which a radial outflow of air could be supplied. Discs of either 102 or 203 mm diameter were rotated at speeds up to 10,000 rev/min, and the average mass-transfer coefficient, h_m, was determined by weighing the disc before and after testing. For

$0.12 < G < 0.6$, $0 < Re_m < 4\text{x}10^4$ (where $Re_m = C_w/2\pi G$) and $5\text{x}10^3 < Re_\phi < 10^5$, the average Sherwood number, Sh_{av}, was correlated by

$$Sh_{av} = (\tfrac{1}{2}G)^{0.55} Re_m^{0.83-12R'_\phi} \Gamma(R'_\phi) \qquad (8.07)$$

where $Sh = h_m b/D_v$ (D_v being the diffusion coefficient for napthalene in air), $R'_\phi = Re_\phi/10^5$ and

$$\Gamma(R'_\phi) = 1.36 + 1.29R'_\phi + 3.57R'^2_\phi - 3.51R'^3_\phi + 1.84R'^4_\phi . \qquad (8.08)$$

The Schmidt number, $Sc = \mu/\rho D_v$, was 2.4, and so equation (8.07) could be used to calculate the average Nusselt number for an isothermal disc rotating in a fluid with a Prandtl number $Pr = 2.4$. However, owing to the polynomial fit used, it may be inaccurate to extrapolate the results outside the test range.

Kreith *et al* noted that the presence of the stator caused premature transition from laminar to turbulent flow on the rotor, and radial outflow further accelerated the onset of transition. The flow over the entire surface of the rotating disc was found to be turbulent for $Re_m > 5\text{x}10^3$.

Kapinos (1965) solved the turbulent momentum-integral equations for the case of a superposed radial outflow between a rotor and stator in which the axial gap was wide enough for core flow to exist. He assumed potential flow in the core: the radial component of velocity was invariant with z and the tangential component was taken to be zero. (These assumptions about conditions in the core are not appropriate to most practical situations.) Using the Reynolds analogy, the calculated Nusselt numbers for $C_w = 0$ were equal to those for a free disc, and for $C_w \rightarrow \infty$ the Nusselt number became independent of Re_ϕ.

Kapinos compared his theoretical results with average Nusselt numbers measured on an air-cooled rotor of 615 mm diameter, over which the temperature was approximately quadratic. For the ranges $0.37 < x_a < 0.47$ ($x_a = a/b$), $0.016 < G < 0.065$, $0.6 < K_0 < 7$, $5\text{x}10^5 < Re_\phi < 4\text{x}10^6$, where

$$K_0 = \frac{2\pi G Re_\phi x_a^2}{C_w}, \tag{8.09}$$

the results were correlated by

$$Nu_{av} = 0.0346\, x_a^{0.3}\, G^{0.06} K_0^{-0.1} Re_\phi^{0.8}, \tag{8.10}$$

where the definition (8.01) is used for the average Nusselt number. It should be noted that the exponent for K_0 was determined from the data with $x_a = 0.37$, and the exponent for x_a was found for $K_0 = 2$ and with $x_a = 0.37$ and 0.47. It follows that, like all correlations, equation (8.10) should be used with caution.

Whereas equation (8.10) implies that the effect of C_w is weak, equation (8.07) suggests that it is strong. These correlations are valid over different ranges of the relevant dimensionless parameters, and they represent two extremes. In fact, Kapinos' solution of the integral equations indicates that $Nu_{av} \rightarrow Nu_{av,fd}$ as $K_0 \rightarrow \infty$ and that $Nu_{av} \propto (C_w/G)^{0.8}$ as $K_0 \rightarrow 0$. This behaviour was also observed, theoretically and experimentally, by Owen, Haynes and Bayley (1974).

For their experiments, Owen *et al* used apparatus similar to that described in Section 7.4.2. Cooling air was supplied through a hole of 102 mm diameter in the centre of the (adiabatic) stator, and an (approximately) quadratic temperature profile was created on the rotor by means of stationary radiant heaters. The heat transfer was calculated in two separate ways: from a "conduction solution" (solving Laplace's equation for the rotor); and from an "enthalpy balance" (using the measured temperatures of the cooling air at inlet and outlet, together with the measured windage torque of the rotor, to calculate the heat transfer). The average Nusselt numbers were defined using equation (8.04).

As well as the experimental results, numerical solutions of the turbulent boundary-layer equations were obtained using the method of Spalding and Patankar (1967). The measured surface temperatures on the rotor and stator were used as boundary conditions for the energy equation.

For $0.006 < G < 0.18$, the theoretical and experimental variation of Nu^*_{av} with G is shown in Figure 8.1. It can be seen that Nu^*_{av} increases as G decreases for small gap ratios and becomes independent of G for large ones. For the intermediate values of G, Nu^*_{av} (like C_M, as shown in Figure 7.13) is smaller than the free-disc value if $C_w < C_{w,fd}$ (where $C_{w,fd}$, the free-disc entrainment rate, can be estimated from equation (4.17)).

Figure 8.2 shows the variation of Nu^*_{av} with Re_ϕ for $G = 0.03$, and the agreement between the experimental and numerical results is, in the main, good. As discussed above, for large rotational speeds the results tend to the

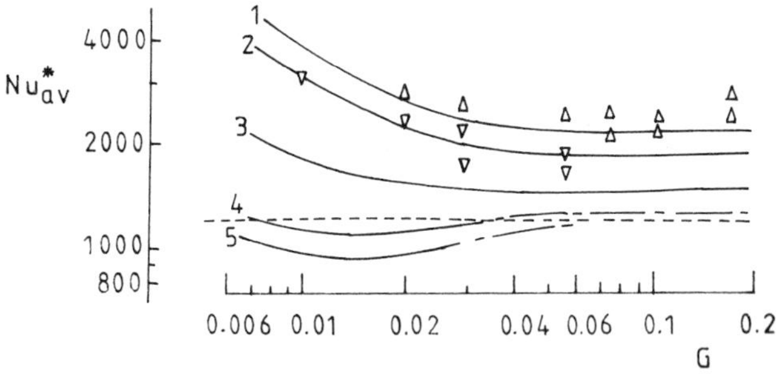

Figure 8.1 Effect of C_w on the variation of Nu^*_{av} with G: $Re_\phi = 9x10^5$ (Owen *et al* 1974)

——— numerical solutions

Curve	1	2	3	4	5
$C_w x10^{-4}$	7.4	4.7	2.2	1.3	0.53

- - - - - free-disc correlation, equation (5.87)

Experiment: △ $C_w = 7.4x10^4$, ▽ $C_w = 4.7x10^4$

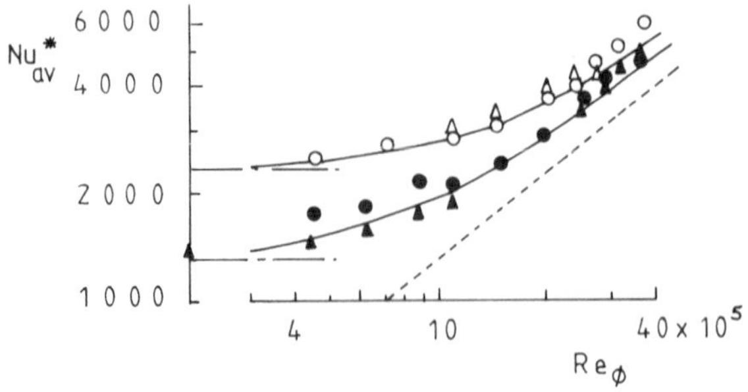

Figure 8.2 Effect of C_w on the variation of Nu^*_{av} with Re_ϕ: $G = 0.03$ (Owen *et al* 1974)

————	numerical solution
— · — · —	asymptotic solution, equation (8.11)
- - - - -	free disc correlation, equation (5.87)

Experiment: △ enthalpy method, ○ conduction method
Open symbols, $C_w = 9.8\text{x}10^4$; Solid symbols, $C_w = 4.7\text{x}10^4$

---

free-disc case (correlated by equation (5.87)), and for small speeds they become virtually independent of Re_ϕ. For the large-flow-rate case ($C_w/GRe_\phi \gg 10$), the Reynolds analogy gave an asymptotic solution in the form

$$Nu^*_{av} = 0.0145\left(\frac{C_w}{G}\right)^{0.8} . \quad (8.11)$$

Although (as discussed in Section 7.4.2) numerical solutions were not obtained for $G > 0.055$, the results obtained at $G = 0.055$ were found to be representative of the values measured at large gap ratios. This is illustrated in Figure 8.3 which shows the variation of Nu^*_{av} with Re_ϕ for $G = 0.08$ and 0.12.

The experimental data of Owen *et al* were correlated by Owen and Haynes (1976) using the modified Reynolds analogy (see equation (2.133)) with $\chi = Pr^{-\frac{2}{5}}$,

$$Nu^*_{av} = \frac{Re_\phi C_M Pr^{0.6}}{\pi} \quad (8.12)$$

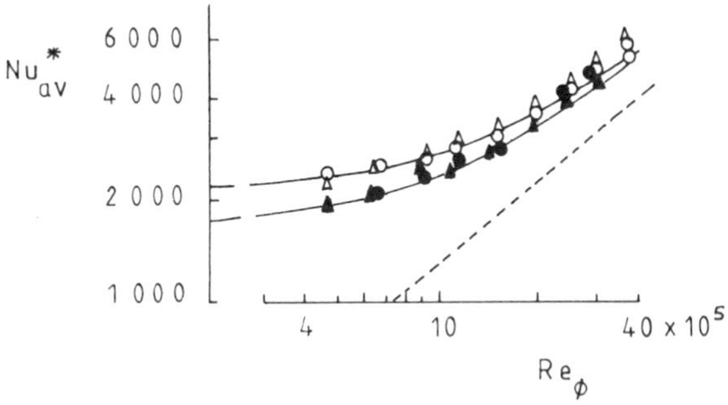

Figure 8.3 Effect of C_w and G on the variation of Nu^*_{av} with Re_ϕ for large gap ratios (Owen *et al* 1974)

——————— numerical solutions

\- - - - - free disc correlation, equation (5.87)

Experiment: △ $G = 0.08$, ○ $G = 0.12$

Open symbols, $C_w = 9.8\text{x}10^4$; Solid symbols, $C_w = 7.4\text{x}10^4$

---

and C_M is their measured moment coefficient. Figure 8.4 shows a comparison between this correlation, the numerical solutions and the experimental data for $G = 0.02$. Although the experimental values tend to be slightly higher at the larger values of Re_ϕ, the overall agreement is good, and it illustrates the asymptotic behaviour referred to above. It should be noted that the correlation is only valid for the same ranges as given for C_M in Section 7.4.2; in particular, it is invalid for $C_w < C_{w,fd}$.

8.2.3 Shrouded systems with radial outflow

Haynes and Owen (1975) extended the radial-outflow work on the shrouded system described in Section 7.4.3 to the heat-transfer case. The shroud was modelled numerically using a modified geometry, and the boundary-layer energy equation was solved (for $G = 0.055$) using the measured temperatures as boundary conditions. Figure 8.5 shows the variation of Nu^*_{av} with Re_ϕ for $G = 0.06$, and it can be seen

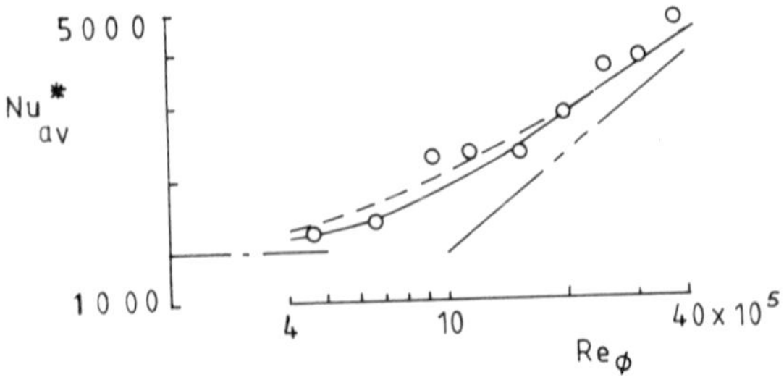

(a) $C_w = 3.2\text{x}10^4$

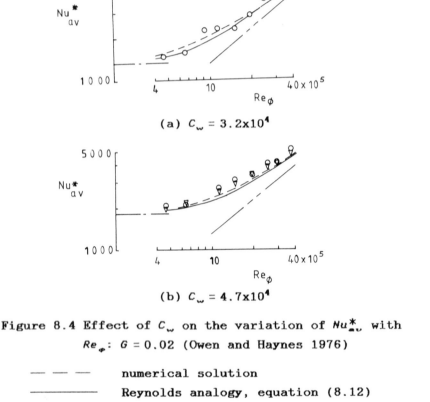

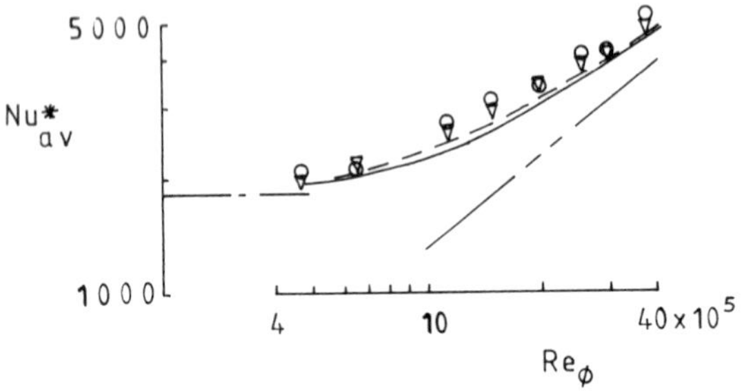

(b) $C_w = 4.7\text{x}10^4$

Figure 8.4 Effect of C_w on the variation of Nu^*_{av} with Re_ϕ: $G = 0.02$ (Owen and Haynes 1976)

— — — numerical solution

———— Reynolds analogy, equation (8.12)

— · — · – asymptotic solution, equation (8.11)

\- - - - - free disc correlation, equation (5.87)

Experiment: o conduction method, ∇ enthalpy method

---

that the numerical model accounts quite accurately for the increase of Nu^*_{av} with decreasing G_c and with increasing C_w. The flow rate was large enough to ensure that Nu^*_{av} was greater than the free-disc case: for large values of Re_ϕ, the results tend towards those for the free disc; for small values of Re_ϕ, the Nusselt numbers are virtually independent of rotational speed. Similar results were also obtained for the unshrouded systems described above.

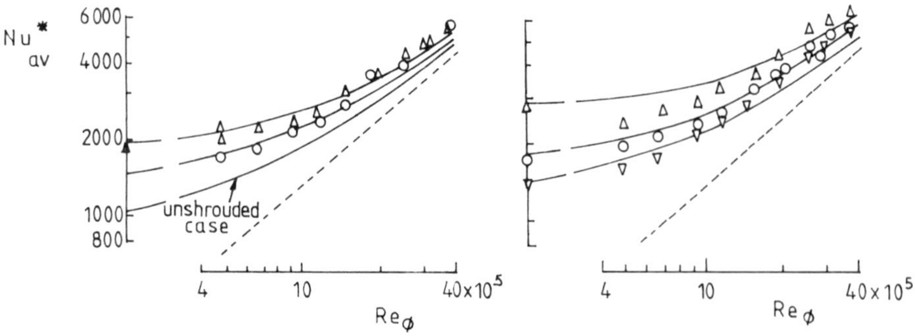

(a) $C_w = 4.7\text{x}10^4$, (b) $G_c = 0.0033$

Figure 8.5 Variation of Nu^*_{av} with Re_ϕ for a shrouded system: $G = 0.06$ (Haynes and Owen 1975)

———— Numerical solution

- - - - Free disc correlation, equation (5.87)

Experiment: (a) △ $G_c = 0.0033$, ○ $G_c = 0.0067$

(b) △ $C_w = 7.5\text{x}10^4$, ○ $C_w = 4.8\text{x}10^4$, ▽ $C_w = 3.1\text{x}10^4$

Figure 8.6 shows the variation of Nu^*_{av} with G_c for $G = 0.12$ and $C_w = 4.7\text{x}10^4$. It can be seen that, in the main, the numerical solutions agree closely with the experimental data. As $G_c \to G$, the results tend to those for the unshrouded system.

Sparrow and his co-workers carried out a number of experimental investigations on the geometries shown in Figure 8.7. Yu, Sparrow and Eckert (1973) conducted experiments for the case shown in Figure 8.7a where the diameters of the rotor, stator and shroud were 451, 432 and 457 mm respectively, and the axial spacing between the rotor and the stator was adjustable. The rotor and shroud were heated electrically and the stator was thermally insulated. Air entered the system through a hole of 51 mm diameter in the hollow rotor shaft and left via a 13 mm clearance between the stationary shroud and the stator.

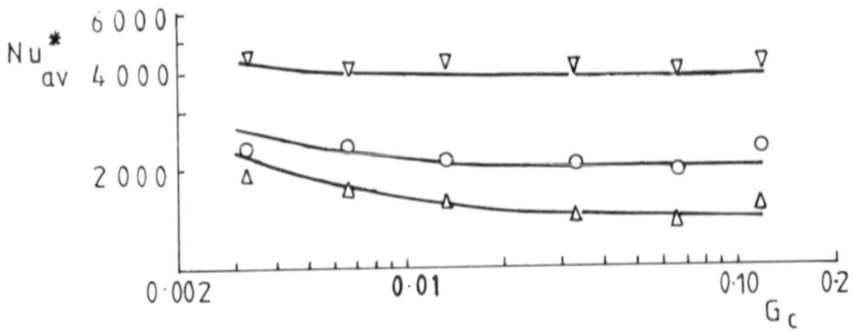

Figure 8.6 Effect of G_c on Nu^*_{av}: $G = 0.12$, $C_w = 4.7\times10^4$ (Haynes and Owen 1975)

——— numerical solution

Experiment:

△ $Re_\phi = 4.8\times10^5$, ○ $Re_\phi = 9.8\times10^5$, ▽ $Re_\phi = 2.7\times10^6$

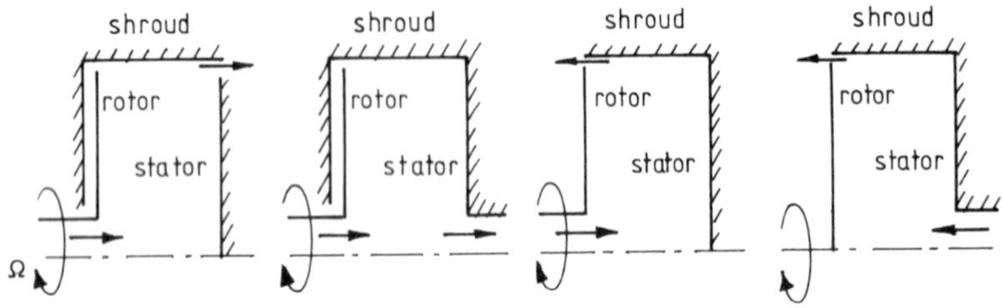

Figure 8.7 Shrouded geometries used by Sparrow and his co-workers

---

Flow visualization, using smoke injected through the shroud, was conducted by Yu *et al* for $0.44 \le G \le 2$, $2.7 \le C_w/10^3 \le 8.8$ and $0.44 \le Re_\phi/10^6 \le 1$. For small ratios of flow rate to rotational speed, the flow was axisymmetric with radial outflow on the rotor and inflow on the stator. At large ratios, however, the flow became nonaxisymmetric: the coolant entered the system at an acute angle to the central axis of the rotor, and the resulting jet precessed about this axis at a rotational speed less than that of the rotor. This phenomenon is believed to be related to the vortex breakdown observed by Owen and Pincombe (1979) for a rotating cavity with a radial outflow of air. However, vortex breakdown was not observed for the rotor-stator tests reported above by Haynes and Owen; in their system (in contrast to those of Yu *et al* and of Owen and Pincombe) the flow entered through a *stationary* pipe at the centre of the stator.

Yu *et al* also determined local and average Nusselt numbers for the heated rotor and shroud using the measured power input to evaluate the heat fluxes. Correlations of Nu for the rotor were obtained for both laminar and turbulent flow (with transition assumed when $x^2 Re_\phi$ was 1.5×10^5). Nu was found to increase as Re_ϕ and C_w increased and as G decreased.

Sparrow, Shamsundar and Eckert (1973) made measurements in the system shown in Figure 8.7b, and, like Yu *et al*, they also observed vortex breakdown. Sparrow, Buszkiewicz and Eckert (1975) and Sparrow and Goldstein (1976) made heat transfer measurements in the systems shown in Figures 8.7c and 8.7d respectively. Numerical analyses of these, or related, systems were carried out by Hennecke, Sparrow and Eckert (1971) for laminar flow and by de Socio, Sparrow and Eckert (1976) for turbulent flow.

Gosman, Lockwood and Loughhead (1976), using the numerical technique outlined in Section 7.4.3, obtained reasonable predictions of the local Nusselt numbers measured by Yu *et al* and of the average Nusselt numbers measured by Haynes and Owen.

8.2.4 Radial inflow

Heat transfer for a rotor-stator system with a radial inflow of air was investigated by Mitchell and Metzger (1965). The rotor, which was 406 mm diameter, was radiantly heated, and the local heat fluxes and temperatures were measured at five locations by embedded fluxmeters and thermocouples. The cooling air entered the system through a peripheral porous shroud, which was attached to the rotor, and left, via stationary vanes, through a hole in the centre of the stator. For $G = 0.110$ and 0.178, $2 < 10^{-5}|C_w|/G < 6$, $0.3 < GRe_\phi/|C_w| < 1.3$ and $Pr = 0.71$, the data were correlated by

$$Nu_{av} = 7.46\left(\frac{|C_w|}{G}\right)^{0.35}\left[1 + 0.45\frac{Re_\phi G}{|C_w|}\right]. \tag{8.13}$$

For these flow-rates, which are greater than the free-disc entrainment rate, the Nusselt numbers are only weakly affected by rotational speed.

To simulate film cooling on a radial turbine, Metzger and Mitchell (1966) extended the experiments to the case where there were two inflows: the cold main inflow, as before, and a hot secondary inflow (that is, film heating rather than cooling) injected adjacent to the rotor. The correlation given by equation (8.13) was modified to take account of the thermal ratio and mass ratio of the two flows.

8.3 IMPINGING JETS

8.3.1 Effect of impingement on heat transfer

Metzger and his co-workers conducted experimental studies of the effect of jet impingement on heat transfer from a rotating disc. Metzger and Grochowsky (1977) obtained average heat transfer rates for tapered copper discs of 76 and 127 mm diameter rotating in air. A nozzle, with its axis coincident with or parallel to the axis of rotation of the disc, was used to inject a jet of air towards the surface of the rotating disc. The disc was heated to 90° C when it was stationary, and it was then accelerated to the test speed after which the cooling rate, measured by an infrared detector, was used to determine the heat loss.

From tests conducted using nozzles of different diameters, d, and different axial distances, h, from the disc, it was concluded that the spacing effect was not significant for $1 < h/d < 5$. Flow visualization revealed an impingement-dominated and a rotation-dominated regime. For the former regime, the jet penetrates the boundary layer on the rotating disc and flows radially inward as well as outward; for the latter regime, there is little penetration and little flow radially inward. The transition between the regimes could be abrupt, and it was related to the ratio of the flow-rate of the jet, $\dot{m}_j$, to the free-disc entrainment rate. Most of the tests were conducted in the laminar or transitional Reynolds-number range, $Re_\varphi < 3\text{x}10^5$, where the appropriate free-disc entrainment rate is given by equation (3.28). For $10^4 \le Re_\varphi \le 3\text{x}10^5$, transition from impingement-dominated to rotation-dominated regimes occurred for $0.05 \le \dot{m}_j/\dot{m}_{fd} \le 1$. In the rotation-dominated regime, the Nusselt numbers were virtually unaffected by the jet; in the impingement-dominated regime, Nu_{av} increased with increasing flow-rate.

Metzger, Mathis and Grochowsky (1979) examined the effects of disc geometry and of impingement cooling on the heat transfer at the rim of the disc. A cross-section of the disc-geometry is shown in Figure 8.8: faces A to D were tested (one at a time) on the rig. The overall diameter of the rotor was 203 mm, and the nozzle was placed at a radial distance of 92 mm from the central axis and an axial distance of 6.4 mm from the disc surface. Most tests were conducted using a single nozzle, with a jet diameter of either 1.59 or 2.38 mm, but some tests were made using two nozzles with jet diameters of 2.38 mm. The rim of the disc was made from aluminium and the shaped faces and hubs from acrylic. Transient tests were conducted and, after corrections were made for conduction losses, the heat

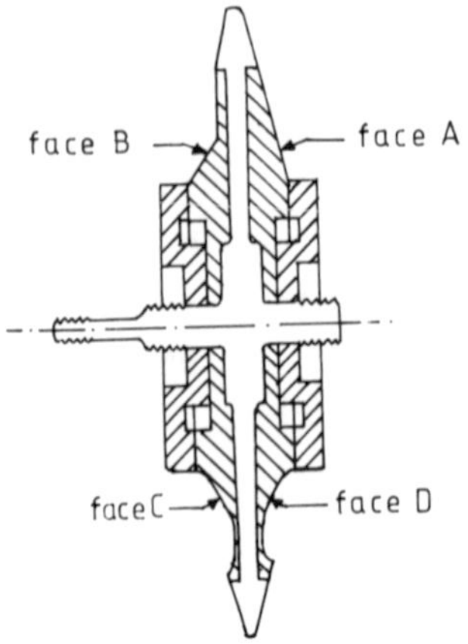

Figure 8.8
Cross-section of disc geometry
(Metzger *et al* 1979)

transfer coefficients from the rim were calculated to an estimated accuracy of 12 to 30% depending on conditions.

Without impingement, the disc geometry had little effect on heat transfer from the rim, which behaved like a free disc. The effect of impingement on the Nusselt number is shown in Figure 8.9 for $Re_\phi = 2.43\text{x}10^5$ (the highest value tested). It should be noted that Nu and Re_ϕ were calculated using the area-weighted average radius of the rim, $\dot{m}_j$ is the mass flow-rate through each jet, and $\dot{m}_e$ is a calculated entrainment rate. (For the rotor used, Metzger *et al* calculated $\dot{m}_e$ to be about 80% of that given by equation (4.17) for a turbulent free disc.)

From Figure 8.9, it can be seen that for $\dot{m}_j/\dot{m}_e < 0.1$ the impingement has little effect, whereas for $\dot{m}_j/\dot{m}_e \simeq 1$, Nu is approximately doubled. It can also be seen that the effects of surface geometry are relatively small. From the limited number of tests conducted with two nozzles, it was observed that in the impingement-dominated regime Nu could be twice the value achieved by a single nozzle. These effects were also apparent for the tests at lower Reynolds numbers, and the authors concluded that rim cooling would be effective only if at least 10% of the disc-pumping flow were supplied by each nozzle.

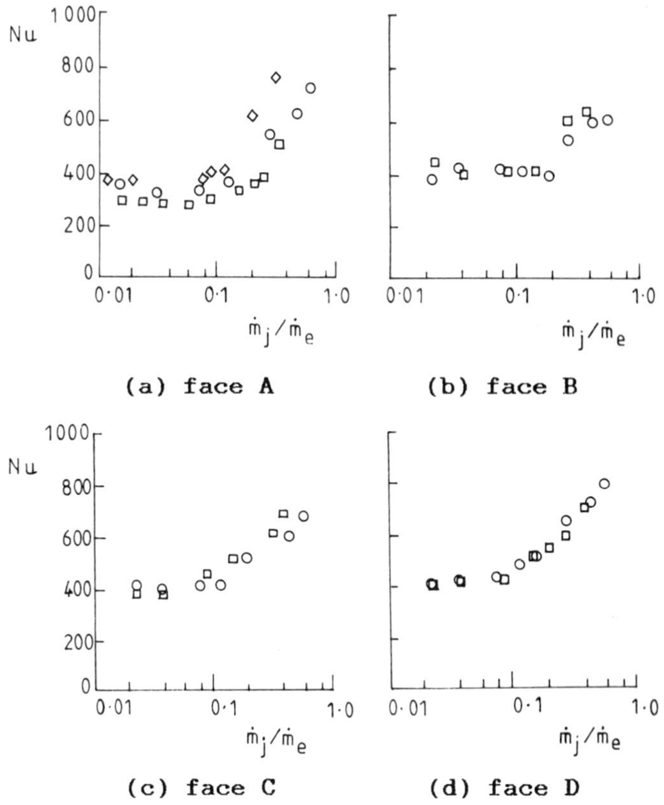

Figure 8.9 Variation of *Nu* with $\dot{m}_j/\dot{m}_e$: $Re_\phi = 2.43 \times 10^5$
(Metzger *et al* 1979)

○ one nozzle (d = 2.38mm)

□ one nozzle (d = 1.59mm)

◇ two nozzles (d = 2.38mm)

Carper, Saavedra and Suwanprateep (1986) made heat-transfer measurements for a horizontal heated disc, rotating in air, cooled by a laminar vertical jet of oil. Tests were carried out, using discs and jets of different diameters, for a range of rotational speeds, flow-rates and Prandtl numbers. The disc was heated electrically and, from the measured power input and temperature differences, the average Nusselt number was evaluated to an estimated accuracy of ± 13%.

The authors observed that the disc/jet diameter ratio, D/d, had no effect on heat transfer for $12.5 < D/d < 141.5$. For the case where the jet impinged on the centre of the disc, their results were correlated for $87 < Pr < 400$, $180 < Re_j < 1300$ and $4 \times 10^3 < Re_\phi < 1.36 \times 10^5$ by

$$Nu_{av} = 0.083\, Pr^{0.448} Re_j^{0.59} Re_\phi^{0.384}, \qquad (8.14)$$

where Re_j is the Reynolds number of the jet (based on its diameter and axial velocity). To minimize variations in the properties of the oil (on which the dimensionless groups were based), the temperature difference between the jet and the disc was kept to approximately 3° C. Correlations were also obtained for the case where the axis of the jet was displaced from that of the disc, but the change in Nu_{av} was found to be relatively small.

8.3.2 Influence of pre-swirl on cooling effectiveness

In many gas turbines, the cooling air for the turbine blades is supplied from stationary "pre-swirl" nozzles. The nozzles impart a tangential component of velocity, in the direction of the rotation of the disc, so that the temperature of the cooling air is reduced relative to the rotating blades (see Meierhofer and Franklin 1981).

A simplified representation is shown in Figure 8.10. Cooling air from the pre-swirl nozzles finds its way into the blade-cooling holes in the rotor, and from there it flows through the cooling passages inside the turbine blades attached to the periphery of the rotor. A question of interest to the designer is: knowing the temperature of the disc-cooling and pre-swirl air at inlet to the system, what is the temperature of the blade-cooling air?

Although the actual flow is three-dimensional, El-Oun

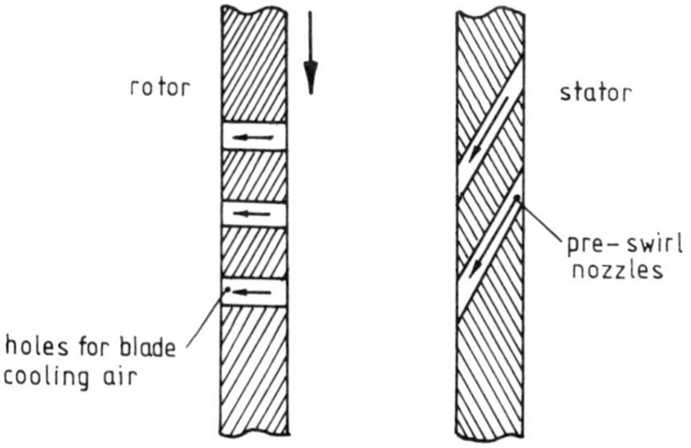

Figure 8.10
Symbolic diagram of pre-swirl nozzles

and Owen (1988) estimated the average effects for a system in which heat transfer from the rotor to the cooling air was negligible. For the case where separate boundary layers exist, equation (2.120) can be used to calculate the adiabatic-disc temperature, $T_{0,ad}$, such that

$$T_{0,ad} = T_{\infty,ad} + R\frac{\Omega^2 r_p^2}{2c_p}(1 - \beta)^2, \qquad (8.15)$$

where r_p is the radial location of the pre-swirl nozzles and of the blade-cooling holes, and $\beta\Omega r$ is the relative rotational speed of the fluid outside the boundary layer on the rotor. It is assumed that the recovery factor, R, could be approximated for air by $R \simeq Pr^{\frac{1}{3}}$. The simplest hypothesis to make is that $T_{\infty,ad}$ and β refer to conditions at the outlet from the nozzles and that the temperature of the blade-cooling air is equal to $T_{0,ad}$. That is

$$T_b = T_{0,ad}, \quad T_{\infty,ad} = T_p, \quad \beta = \frac{v_{\phi,p}}{\Omega r_p}, \qquad (8.16)$$

where the subscripts b and p refer to the blade-cooling and pre-swirl flows respectively.

In the nozzles the total enthalpy, H_p is conserved where

$$H_p = c_p T_p + \tfrac{1}{2}(v_r^2 + v_\phi^2 + v_z^2)_p, \qquad (8.17)$$

so that T_p decreases as $v_{\phi,p}$ increases. It is convenient, therefore, to define a nondimensional temperature difference, $\Theta(\beta)$, as

$$\Theta(\beta) = \frac{H_p - c_p T_{0,md}}{\frac{1}{2}\Omega^2 r_p^2} . \tag{8.18}$$

Θ provides a measure of the pre-swirl cooling-effectiveness in terms of the difference between the total enthalpy of the air leaving the nozzles and that reaching the rotor (and hence the blades). It follows from equations (8.15) and (8.18) (assuming $v_{r,p}^2$, $v_{z,p}^2 \ll v_{\phi,p}^2$) that

$$\Theta(\beta) = \beta^2 - R(1 - \beta)^2 . \tag{8.19}$$

Increasing β therefore increases Θ, thereby decreasing the temperature of the blade-cooling air. (In an engine, the available pressure drop limits the pre-swirl that can be imparted to the cooling air.)

In practice, the air from the pre-swirl nozzles can mix with the disc coolant: the blade-cooling air becomes "thermally contaminated". To allow for this, El-Oun and Owen used a simple model which assumed "perfect mixing" between the pre-swirl and disc-cooling flows. The "effective value" of β was calculated from the conservation of angular momentum for the two flows, and the "effective value" of H_p was calculated from the steady-flow energy equation (including the work put into the system by the frictional moment on the rotor, calculated using the method described in Section 7.3.2). These effective values can then be used in equations (8.19) and (8.15) to calculate Θ and $T_{0,md}$.

A schematic diagram of the rig is shown in Figure 8.11; $C_{w,b}$, $C_{w,d}$, $C_{w,p}$ and $C_{w,s}$ refer to the blade-cooling, disc-cooling, pre-swirl and sealing flows respectively. The diameter of the rotor was 442 mm, and the radius of the centre-line of the pre-swirl nozzles was 200 mm

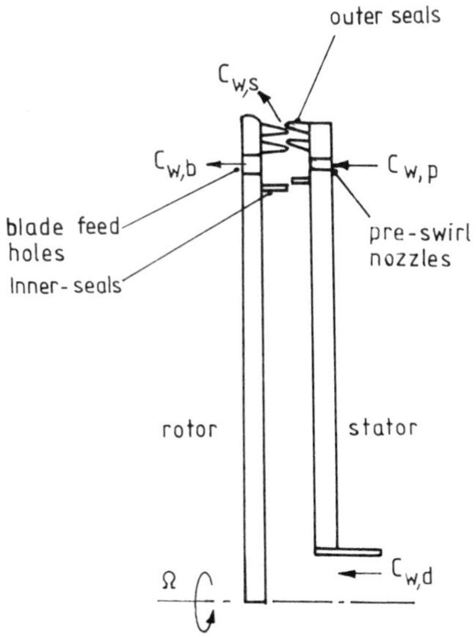

Figure 8.11
Pre-swirl rig

($r_p/b \simeq 0.9$). The axial spacing between the rotor and stator was 24 mm ($G \simeq 0.11$), and there were 60 nozzles of 2.9 mm diameter, inclined at an angle of 70° to the axial direction, in the stator and 60 holes of 7.7 mm diameter in the rotor. The temperature of the blade-cooling air was measured by total-temperature probes located in five of the holes in the rotor, and the thermocouple signals were brought out via a slipring assembly; the other air temperatures were measured by stationary total-temperature probes. The radial clearance of the outer (double-mitre) seals was set at either 1.1 or 1.6 mm, and the clearance of the inner seal (which was not fitted for all tests) was 0.7 mm.

Figure 8.12 show the variation of Θ with the *swirl ratio* s_r (whère $s_r = v_{\phi,p}/\Omega r_p$ is calculated from conditions at the pre-swirl nozzles) for radial outflow and inflow; for a fixed value of $C_{w,p}$, s_r was varied by altering the speed of the rotor. The "theory for unmixed flows" assumes that $\beta = s_r$ in equation (8.19); the "theory for mixed

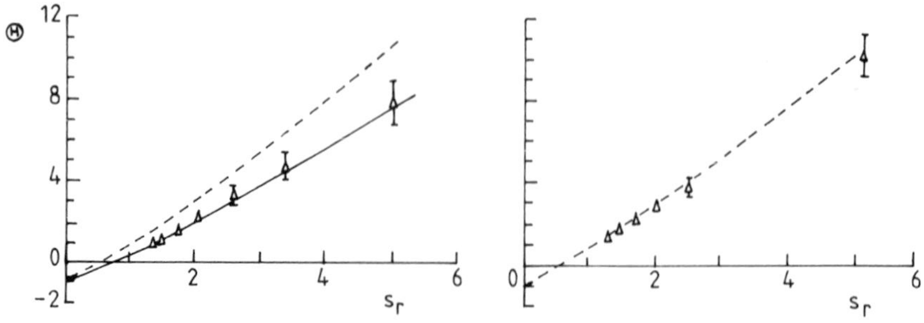

(a) Radial outflow: $C_{w,b} = 9500$ $C_{w,d} = 5500$

(b) Radial inflow: $C_{w,b} = 4050$ $C_{w,d} = -2400$

Figure 8.12 Variation of Θ with s_r: $C_{w,p} = 1.8\text{x}10^4$ (El-Oun and Owen 1988)

- - - - - "theory for unmixed flows"

———— "theory for mixed flows"

experiment and error bounds

---

flows" uses the mixing model outlined above.

It can be seen from Figure 8.12a that, for radial outflow, the mixed-flow theory agrees very well with the majority of the experimental data. For radial inflow with $s_r > 1$, the flow can move radially inward on both the rotor and the stator, and "thermal contamination" of the blade-cooling flow should not occur. Under these conditions, the unmixed-flow theory is appropriate, as shown in Figure 8.12b. For these results, the inner seals shown in Figure 8.11 were not fitted; for other tests, with both radial outflow and inflow, it was found that the inner seals had no significant effect on the measured values of Θ.

CHAPTER 9

Sealing Rotor-Stator Systems: The Ingress Problem

A simple shrouded rotor-stator system is shown in Figure 2.1. At sufficiently high rotational speeds, Batchelor-type flow (see Section 6.2.1) occurs: separate boundary layers form on the rotor and stator and a core of rotating fluid is formed. The rotating core creates an adverse pressure gradient ($\partial p/\partial r > 0$), and the pressure inside the wheel-space between the rotor and the stator can become negative (that is, the pressure inside is less than that of the external fluid). Under these conditions ingress occurs: external fluid flows into the wheel-space through the clearance between the shroud and the rotor. The ingested fluid flows down the stator, is entrained into the boundary layer on the rotor via the core, and is then "pumped" out of the system.

In a gas-turbine engine, the ingress of hot mainstream gas into the wheel-space is undesirable as it can cause overheating of the rotor. In practice, ingress is prevented, or at least reduced, by supplying a radial outflow of cooling air. This air, which also removes heat from the turbine disc, tends to "pressurize" the system: a pressure drop occurs as the coolant passes through the small clearances in the seals at the periphery of the system. The designer has to optimize the amount of air used: too little can cause overheating; too much is expensive as the work put into the air by the compressor must be paid for.

Ingress is affected by the geometry of the system and by the speed of the rotor. In addition, it depends on the conditions that occur in the external flow outside the system. These effects are discussed separately below.

9.1 AXIAL-CLEARANCE SEAL IN A QUIESCENT ENVIRONMENT

The fluid dynamics of a shrouded rotor-stator system with an axial-clearance seal, as illustrated in Figure 7.15, was studied by Bayley and Owen (1970). The apparatus was described in Sections 7.4.2 and 7.4.3, and the experiments were conducted in a quiescent environment: the air outside the system was stationary.

Figure 9.1 shows the observed effect of shroud-clearance ratio, G_c, on the measured radial distribution of a pressure coefficient, C_p'. Here

$$G_c = \frac{s_c}{b} \tag{9.01}$$

where s_c is the shroud clearance, and

$$C_p' = \frac{(p - p_\infty)\rho b^2}{\mu^2}, \tag{9.02}$$

where p is the local static pressure (measured by pressure taps on the stator) and p_∞ is the ambient pressure outside the system.

For a constant flow-rate ($C_w = 2.4\text{x}10^4$) and a single gap ratio ($G = 0.06$), it can be seen from Figure 9.1 that the pressure inside the system increases as G_c and Re_ϕ decrease. However, for $x \leqslant 0.8$, the radial pressure *gradient* is not affected significantly by G_c. This suggests that the flow is essentially "parabolic" for $x \leqslant 0.8$, and "elliptic" effects, due to the presence of the shroud, are important only near the outer part of the system. In fact it is possible to obtain boundary-layer solutions for the flow and heat transfer in a shrouded rotor-stator system by assuming a modified geometry near the shroud (see Section (7.4.3).

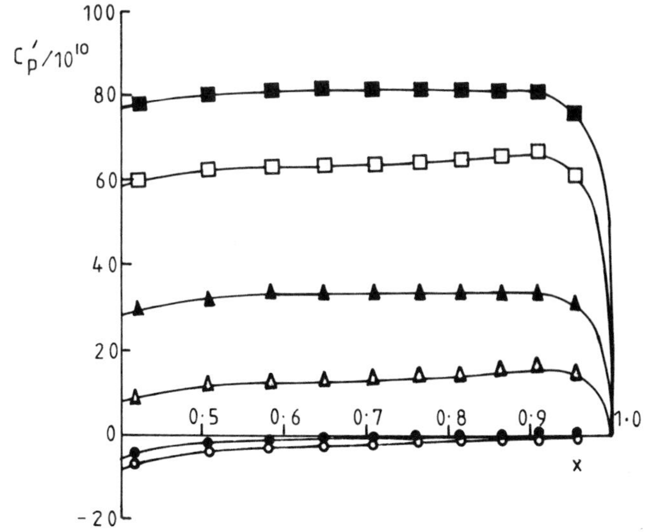

Figure 9.1 Effect of G_c and Re_ϕ on the radial variation of C_p' for an axial-clearance seal: $G = 0.06$, $C_w = 2.4 \times 10^4$ (Bayley and Owen 1970)

○ no shroud, △ $G_c = 0.0067$, □ $G_c = 0.0033$
Open symbols, $Re_\phi = 3.6 \times 10^6$; Solid symbols, $Re_\phi = 0$

It seems reasonable, therefore, to treat the effect of the shroud by superposition: the nondimensional pressure drop across the shroud ($\Delta C_p'$, say) can be added to the distribution occurring at $x \leq 0.8$. The variation of $\Delta C_p'$ (where $\Delta C_p'$ is taken here as the measured value of C_p' at $x = 0.864$) with C_w^2 for $G_c = 0.0067$ is shown in Figure 9.2, and the variation of $\Delta C_p'$ with Re_ϕ^2 is shown in Figure 9.3.

For small values of s_c/s (that is, $G_c/G \ll 1$), the changes in pressure and momentum of the fluid as it leaves the system are likely to be large compared with viscous effects. If $v_{r,c}$ and $v_{\phi,c}$ are the effective values of the radial and tangential components of velocity, where $v_{r,c} = v_{r,c}(r)$ and $v_{\phi,c} = v_{\phi,c}(r)$, then the pressure gradient can be calculated by

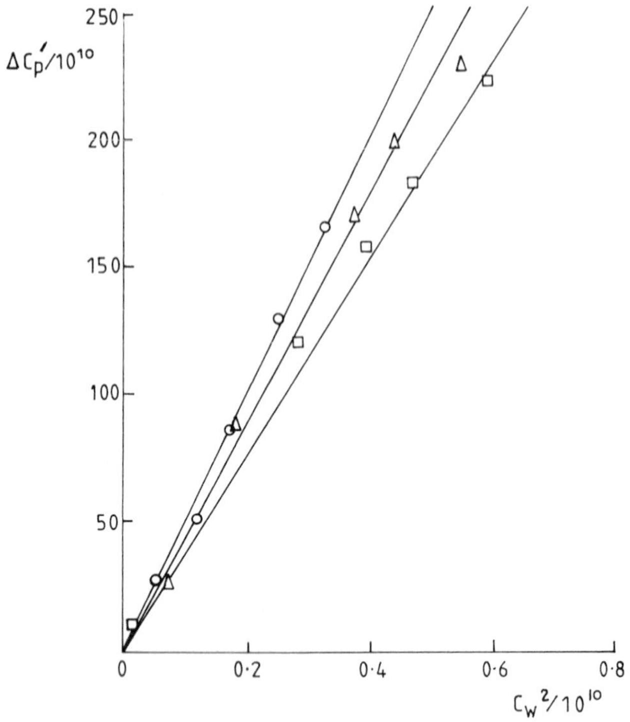

Figure 9.2 Effect of G on the variation of $\Delta C_p'$ with C_w^2 for an axial-clearance seal: $G_c = 0.0067$, $Re_\phi = 0$ (Bayley and Owen 1970)

○ G = 0.06, △ G = 0.12, □ G = 0.18

$$-\frac{1}{\rho}\frac{dp}{dr} = v_{r,c}\frac{dv_{r,c}}{dr} - \frac{v_{\phi,c}^2}{r}. \tag{9.03}$$

Bayley and Owen assumed that, for a mass flow-rate $\dot{m}$,

$$v_{r,c} = \frac{k_1\dot{m}}{2\pi\rho r s'}, \qquad v_{\phi,c} = k_2\Omega r \tag{9.04}$$

where k_1 and k_2 are invariant with r, $s' = s_c$ at $r = b$ and $s' = s$ at $r = b_1$ (b_1 corresponding to the radius at which $\Delta C_p'$ is measured). Hence, integrating equation (9.03) between $r = b_1$ and $r = b$ gives

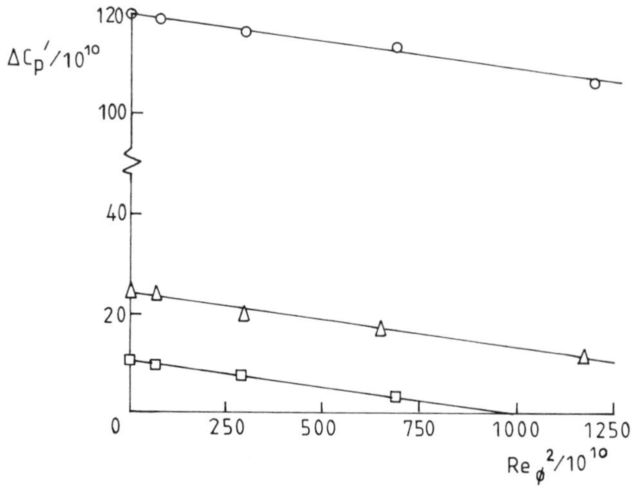

Figure 9.3 Effect of C_w on the variation of $\Delta C_p'$ with Re_ϕ^2 for an axial-clearance seal: $G = 0.18$, $G_c = 0.0067$ (Bayley and Owen 1970)

○ $C_w = 5.4\text{x}10^4$, △ $C_w = 2.4\text{x}10^4$, □ $C_w = 1.6\text{x}10^4$

$$\Delta C_p' = A\, C_w^2 - B\, Re_\phi^2, \tag{9.05}$$

where

$$A = \frac{k_1^2}{8\pi^2}\left[G_c^{-2} - \left(\frac{b_1}{b}G\right)^{-2}\right], \quad B = \tfrac{1}{2}k_2^2\left[1 - \left(\frac{b_1}{b}\right)^2\right]. \tag{9.06}$$

For $0.06 < G < 0.18$, it was found that values of $0.58 < k_1 < 0.66$ and $0.17 < k_2 < 0.21$ provided a good fit to the data for $G_c = 0.0067$ and $b/b_1 = 0.864$. It follows from equation (9.05) that, for large enough values of Re_ϕ, $\Delta C_p'$ can become negative; the change of sign occurs where $\Delta C_p' = 0$ when

$$C_w = \left(\frac{B}{A}\right)^{\frac{1}{2}} G_c\, Re_\phi. \tag{9.07}$$

A separate study was made to determine $C_{w,min}$, the minimum value of C_w necessary to prevent ingress. For these tests, a sub-atmospheric reading on a pressure tap located on the shroud was interpreted as implying ingress: it was assumed that $C_w = C_{w,min}$ when this pressure was zero. Figure 9.4 shows the variation of $C_{w,min}$ with Re_ϕ obtained in this way. For $Re_\phi < 4 \times 10^6$ and $G = 0.06$, 0.12 and 0.18, there was no significant effect of gap ratio, and the results were correlated for $G_c = 0.0033$ and 0.0067 by

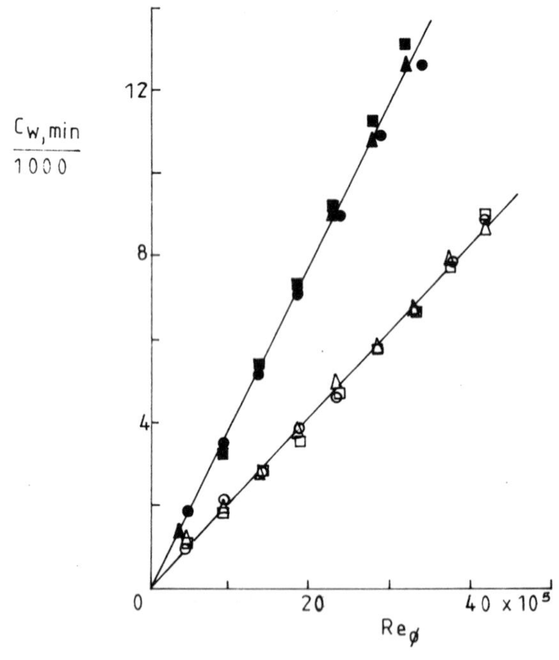

Figure 9.4 Effect of G_c on the variation of $C_{w,min}$ with Re_ϕ for an axial-clearance seal (Bayley and Owen 1970)

○ $G = 0.06$, □ $G = 0.12$, △ $G = 0.18$

Open symbols, $G_c = 0.0033$; Solid symbols, $G_c = 0.067$

$$C_{w,min} = 0.61 G_c Re_\phi. \quad (9.08)$$

This correlation accords with the form given by equation (9.07), but the "constant" of proportionality obtained for the latter, using average values of the relevant quantities, was higher. Although based on a very simple model, equation (9.08) provides a reasonable estimate of $C_{w,min}$ for axial-clearance seals with small values of G_c.

It should be noted that the above model takes no account of the fact that the rotational speed of the core decreases as C_w increases (see Chapter 7). As a consequence, the radial pressure gradient in the wheel-space decreases as C_w increases, and the magnitude of the negative pressure is thereby attenuated. At large values of G_c, where $C_{w,min}$ is high, less flow is required to seal the system than equation (9.08) indicates.

It was shown in Chapter 7 that the core rotation is virtually suppressed when $C_w > C_{w,fd}$, where $C_{w,fd}$ is the free-disc entrainment rate given by equation (4.17) as

$$C_{w,fd} = 0.219 Re_\phi^{0.8}. \quad (9.09)$$

This suggests that equation (9.08) can be valid only for

$$G_c Re_\phi^{-0.2} < 0.36. \quad (9.10)$$

In a gas turbine, the sealing flow-rate is usually less than $C_{w,fd}$ and, unless the clearances are very small, some ingress may occur at the large values of Re_ϕ at which engines operate.

9.2 EFFECT OF SEAL GEOMETRY IN A QUIESCENT ENVIRONMENT

Owen and Phadke (1980), Phadke (1982), and Phadke and Owen (1983, 1988a) examined a number of seal geometries using the seven types of shrouds shown in Figure 9.5. As the results of Bayley and Owen indicated that the effects of gap ratio were weak, Phadke and Owen conducted all

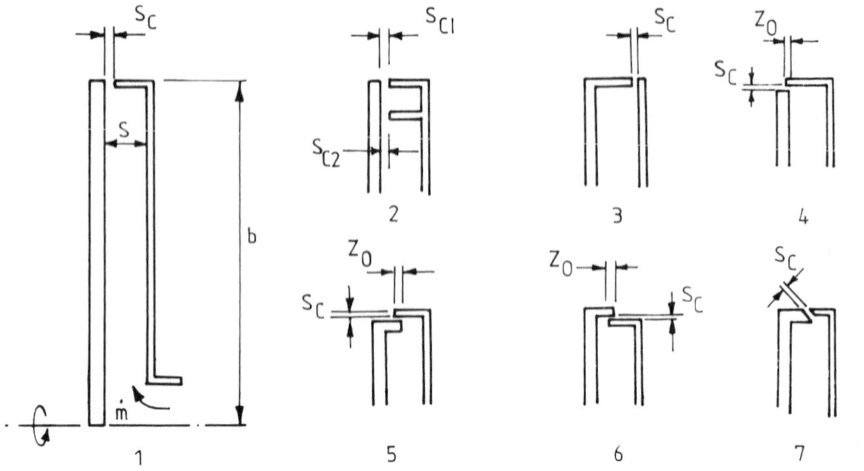

Figure 9.5 Schematic diagram of the seven seal geometries used by Phadke and Owen

---

their experiments with a gap ratio of $G \simeq 0.1$. They used several different rigs, each with a nominal diameter of 380 mm, and the shroud-clearance ratio was varied in the range $0.0025 < G_c < 0.04$ for rotational Reynolds numbers up to $Re_\phi \simeq 1.2 \times 10^6$. Sealing air was supplied through a hole of diameter 38 mm in the centre of the stator, and it left through the seal clearance at the periphery.

The shrouds and the stator were made from acrylic plastic to provide optical access, and the stator was instrumented with pressure taps. Flow visualization was used to observe the flow structure and to determine incipient ingress. A smoke generator released clouds of micron-sized oil particles into the air outside the system, and when ingress occurred smoke was ingested into the wheel-space. $C_{w,min}$ was determined by increasing the flow-rate of sealing air until no smoke was visible inside the system.

An example of the flow visualization is shown in Figure 9.6 for seal 1 with $G_c = 0.01$, $C_w \simeq 400$ and $Re_\phi = 5 \times 10^4$, where the flow is laminar. In Figure 9.6a, the smoke was injected into the sealing air ("central seeding") and the white boundary layers (in which cellular disturbances are present) on the rotor and stator stand out in contrast to the black interior core. The r-z plane of the wheel-space was illuminated by an argon-ion laser, and reflection caused mirror images at the surfaces of the discs. In Figure 9.6b, the smoke was released outside the system ("peripheral seeding"), and the boundary layer on the rotor stands out in black against the white smoke-filled region in the core; increasing the flow-rate would eventually prevent the smoke from penetrating into the wheel-space.

As well as flow-visualization, pressure and concentration measurements were used to determine ingress. The pressure on the stator near the shroud (at $x \simeq 0.97$) was used to determine $C_{w,min}$: when the pressure inside the wheel-space was equal to that outside, the system was assumed to be sealed. Nitrous oxide was ingested into

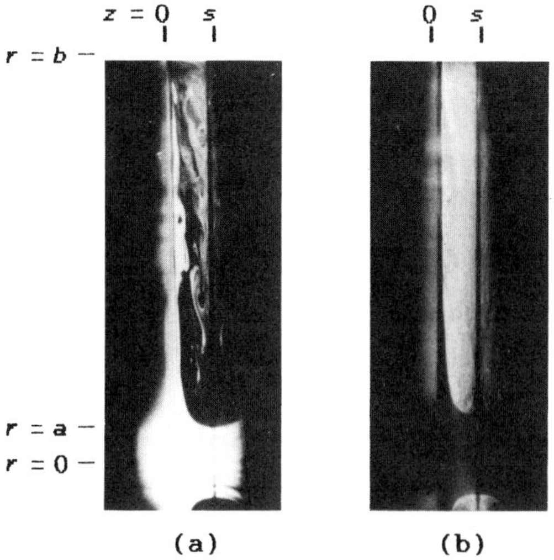

Figure 9.6
The flow structure with seal 1:
$G_c = 0.01$, $C_w = 395$, $Re_\phi = 5\text{x}10^4$
(Phadke 1982)
(a) central seeding
(b) peripheral seeding

(Note: Reflections occur at $z = 0$ and $z = s$)

either the sealing air or the external air, and the measured concentration in the outer part of the wheel-space was used to determine whether or not ingress had occurred. All three techniques yielded results that were qualitatively similar but, not surprisingly, there were quantitative differences.

The overall flow patterns, deduced from flow visualization, are shown in Figure 9.7. These, and other details of the performance of the seven seals, are discussed below.

9.2.1 Axial-clearance seals

Figure 9.8 shows the effect of Re_ϕ and C_w on the radial distribution of p^* (where $p^* = 1000(p - p_\infty)/p_\infty$) for seal 1 with $G_c = 0.01$. As noted in Section 9.1, the pressure in the wheel-space increases as C_w increases and as Re_ϕ

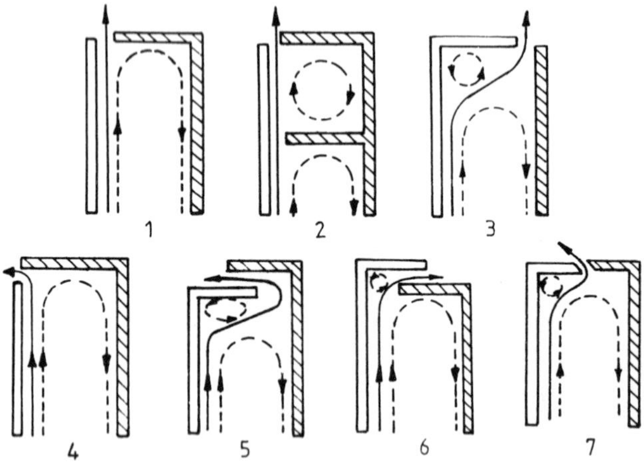

Figure 9.7 Simplified representation of flow patterns near the outlet of each of the seven seals (Phadke and Owen 1988a)

——→—— superposed flow, - - → - - secondary flow

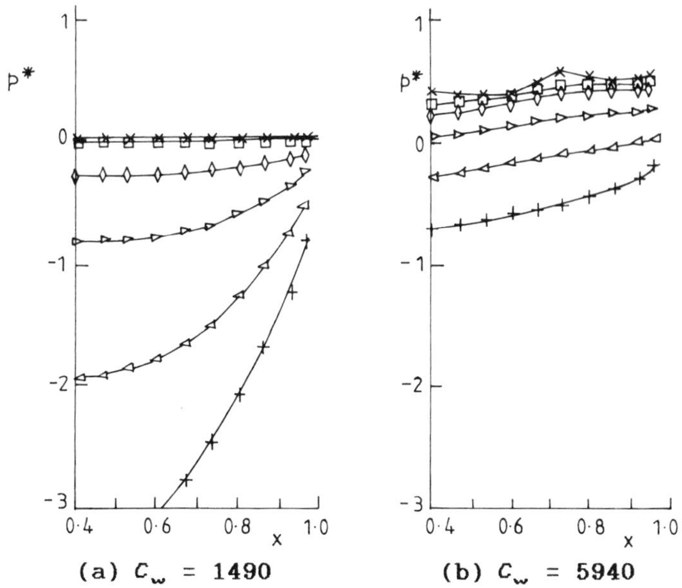

Figure 9.8 Effect of Re_ϕ and C_w on the radial variation of p^* for seal 1 (Phadke and Owen 1988a)

Symbol	X	□	◊	▷	◁	+
Re_ϕ	0	$2x10^5$	$4x10^5$	$6x10^5$	$8x10^5$	10^6

---

decreases. For $C_w = 1490$ and $Re_\phi \geq 4x10^5$, the pressure is negative throughout the wheel-space; for $C_w = 5940$, the pressure near the outer radius is positive at the lower values of Re_ϕ and negative at the higher values. For a sufficiently large value of Re_ϕ, the pressure near the periphery changes from positive to negative; ingress then occurs.

Using both pressure measurements and flow visualization, $C_{w,min}$ was correlated with G_c and Re_ϕ using a multiple-regression analysis based on the assumed form

$$C_{w,min} = c_{i,j}\, G_c^{m_{i,j}}\, Re_\phi^{n_{i,j}}, \tag{9.11}$$

where the constants c, m and n were determined from the analysis. The subscript i refers to the seal number ($i = 1$ to 7), and j refers to the ingress criterion ($j = 1$ for flow visualization, $j = 2$ for pressure measurements). The values of c, m and n are given in Table 9.1 and a comparison between the correlation and the results obtained from flow visualization for seal 1 are shown in Figure 9.9.

The data in Figure 9.9 and subsequent figures were obtained using flow-visualization criteria, unless otherwise stated. For $0.0025 \leq G_c \leq 0.04$ and $0 < Re_\phi \leq 1.2\text{x}10^6$, the correlation for seal 1 is

$$C_{w,min} = 0.280\, G_c^{0.677}\, Re_\phi^{0.956}. \tag{9.12}$$

Also shown is the original Bayley and Owen correlation given by equation (9.08) (which was obtained for $G_c = 0.0033$ and 0.0067 using a pressure criterion). For $G_c = 0.01$, the two correlations are in reasonable agreement; at larger values of G_c the Bayley-Owen criterion overestimates $C_{w,min}$, and at smaller G_c it underestimates. This behaviour is consistent with that outlined in Section 9.1: at large values of G_c, where $C_{w,min}$ is high, the magnitude of the negative pressure created by core rotation is attenuated; equation (9.08), which was based on a simple model, takes no account of this effect. The

---

Table 9.1 Correlation constants (Phadke and Owen 1988a)

flow visualization, $j = 1$; pressure measurements, $j = 2$

Seal i		1	3	4a	4b	5	6	7
c_{ij}	$j = 1$	0.280	0.143	0.091	0.0189	0.0224	0.188	0.042
	$j = 2$	0.197	0.114	0.076	0.0208	0.028	0.149	0.039
m_{ij}	$j = 1$	0.677	0.542	0.482	0.199	0.299	0.620	0.338
	$j = 2$	0.604	0.500	0.473	0.274	0.291	0.580	0.361
n_{ij}	$j = 1$	0.956	0.956	0.950	0.930	0.951	0.951	0.960
	$j = 2$	0.956	0.955	0.950	0.930	0.949	0.950	0.951

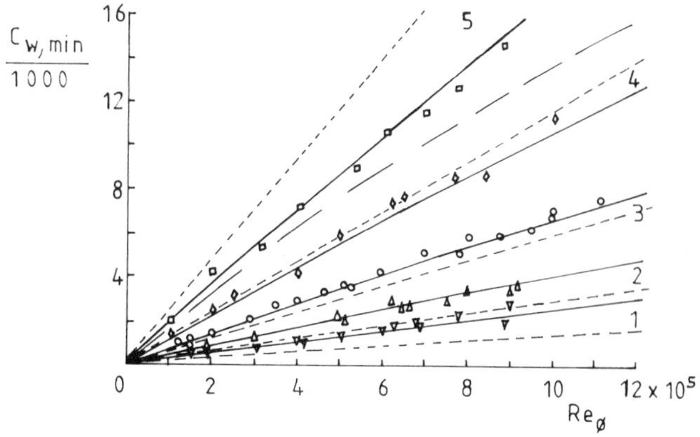

Figure 9.9 Effect of G_c on the variation of $C_{w,min}$ with Re_ϕ for seal 1.

————	correlation (9.12) (Phadke and Owen 1988a)
- - - - -	correlation (9.08) (Bayley and Owen 1970)
— — —	free-disc entrainment rate, equation (9.09)

Symbol	▿	△	○	◇	□
Curve	1	2	3	4	5
G_c	0.0025	0.005	0.01	0.02	0.04

free-disc entrainment curve, equation (9.09), is also shown in the figure, and it can be seen that the results for $G_c = 0.04$ are above the curve; for this value of G_c, the inequality (9.10) indicates that equation (9.08) is invalid.

For the double-shrouded seal (seal 2 in Figure 9.5) the shrouds were made from perspex, 2 mm thick, and pressure taps were fitted on either side of each shroud. The ratio of the radii of the inner and outer shrouds was 0.9, and the clearance ratios for each shroud were independently varied from $G_c = 0.0025$ to 0.01. Ingress was determined from flow visualization and from the pressure difference measured across the taps in the shrouds.

Flow visualization showed that a toroidal vortex formed

in the annulus between the two shrouds, as illustrated in Figure 9.7. For the inner wheel-space, the outer shroud had only a small effect, and $C_{w,min}$ could be correlated with $G_{c,2}$ (where the subscript 2 refers to the inner shroud) and Re_{ϕ}, using the relationships used for seal 1, provided all three nondimensional parameters were based on the inner radius. However, the outer cavity could not be sealed effectively when the clearance of the inner seal was smaller than that of the outer one. Under these conditions, a "sudden-enlargement effect" could cause ingress even when the rotor was stationary! When the clearance of the inner seal was greater than that of the outer one, the behaviour of the outer seal was similar to that observed for the single-seal case discussed above. From these results, it was concluded that (in a quiescent environment) seal 2 had no advantage over seal 1. However, it is shown below that the double-shrouded seal does have some useful characteristics when there is an external flow.

The rotating shroud with axial clearance (seal 3 in Figure 9.5) behaved in a broadly similar way to seal 1. The flow structure (see Figure 9.7) appeared to indicate that the boundary layer on the rotor separated near the periphery, and there were some differences, depending on G_c, between the radial pressure distributions for seals 1 and 3.

The correlation constants for estimating $C_{w,min}$ according to equation (9.11) are given in Table 9.1: the performance of seal 3 is better than that of seal 1 at the larger values of G_c ($G_c \geq 0.01$) and worse at the smaller values. The variation of $C_{w,min}$ with Re_{ϕ}, obtained by flow visualization, is shown in Figure 9.10 for $G_c = 0.01$.

9.2.2 Radial-clearance and mitred-clearance seals

Referring to Figure 9.7, the flow pattern for seal 4 (the

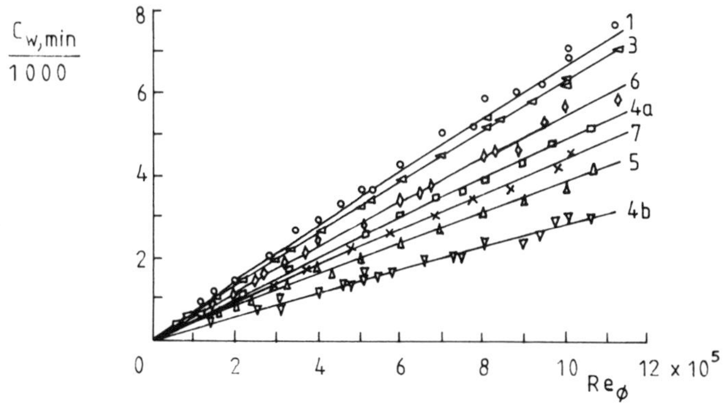

Figure 9.10 Variation of $C_{w,min}$ with Re_ϕ for the seals with $G_c = 0.01$ (Phadke and Owen 1988a)

——— correlation (9.11)

Symbol	○	◁	□	▽	△	◊	×
Seal	1	3	4a	4b	5	6	7

radial-clearance seal in which the stationary shroud overlaps the rotor) was the hardest to deduce. The outward flow on the rotor formed a radial wall jet that impinged on the stationary shroud: the secondary flow travelled across the shroud and moved radially inwards on the stator, and the sealing air left through the radial clearance between the shroud and the rotor. When ingress occurred, the radial jet became nonaxisymmetric and unsteady; when the flow-rate was large enough to prevent ingress, the flow was stabilized. A similar effect occurred for seal 5 (where the stationary shroud overlapped the rotating one); in this case, the flow separated from the rotating shroud and impinged on the stator, as illustrated in Figure 9.7.

The "impinging-jet phenomenon" (which is similar to the "wall-jet effect" discussed in Section 6.5.1) was not seen for seal 6, where the rotating shroud overlapped the stationary one. The flow separated from the rotating shroud in a similar way to that for seal 3, and did not impinge on the stationary surfaces. Seal 7 (the mitred-clearance seal) showed some evidence of impingement, but the effect was not as strong as that for seal 4.

The impinging-jet phenomenon is believed to be the explanation for the "pressure-inversion effect" reported by Phadke and Owen (1983): at large flow-rates, seals 4 and 5 could show an *increase* in pressure with increasing Re_ϕ (rather than the decrease that occurs for the axial-clearance seals). The impinging jet forms a curtain of fluid that creates an increase in pressure in the outer part of the wheel-space. As a consequence, these radial-clearance seals are more effective in preventing ingress than the equivalent axial-clearance ones.

The radial distribution of pressure for seals 1 and 4, with $G_c = 0.005$, is shown in Figure 9.11; seals 4a and 4b refer to the radial-clearance seal 4 with overlap ratios $z_0/b = 0$ and 0.03 respectively (z_0 being the axial overlap). For $C_w = 2950$, seal 4a (like seal 1) shows the "conventional behaviour": p^* decreases as Re_ϕ increases. However, seal 4b shows evidence of the pressure-inversion effect: the pressure near the periphery increases as Re_ϕ increases from 0.8×10^6 to 10^6. This effect is apparent for both seals 4a and 4b for $C_w = 5940$: although all three seals indicate positive pressures throughout the wheel-space, the influence of Re_ϕ is attenuated for the two radial-clearance seals.

Values of $C_{w,min}$ for the radial-clearance and mitred-clearance seals were found to be smaller than for their axial-clearance counterparts operating at equivalent values of G_c and Re_ϕ. The correlation constants for $C_{w,min}$ for seals 4 to 7 are given in Table 9.1, and the variation of $C_{w,min}$ with Re_ϕ for $G_c = 0.01$ is shown in Figure 9.10. It can be seen that seals 4b, 5 and 7 are the

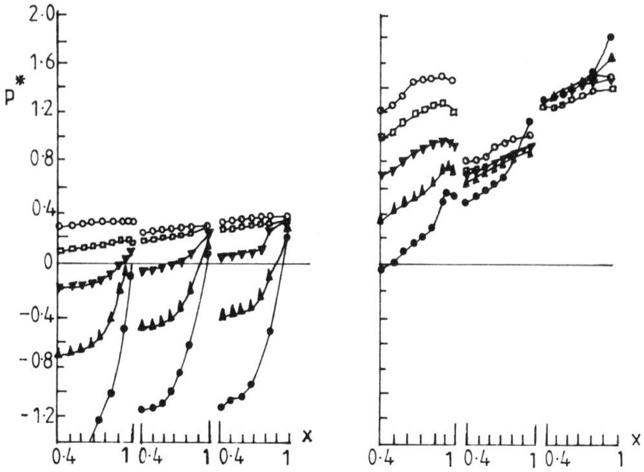

Seal 1 4a 4b 1 4a 4b

(i) C_w = 2950 (ii) C_w = 5940

Figure 9.11 Effect of Re_ϕ and C_w on the radial variation of p^* for seals 1 and 4: $G_c = 0.005$ (Phadke 1982)

Symbol	∘	□	▼	▲	•
Re_ϕ	0	$4\text{x}10^5$	$6\text{x}10^5$	$8\text{x}10^6$	10^6

most effective: they all exhibited the impinging-jet and pressure-inversion effects. In particular, seal 4b requires the smallest amount of sealing air; it also exhibits the most nonlinear behaviour, with the rate of increase of $C_{w,min}$ decreasing as Re_ϕ increases.

The "order-of-merit" for the seals with $G_c = 0.01$ and 0.02 was: seal 4b, 5, 7, 4a, 6, 3 and 1. For $G_c = 0.005$, seal 1 was slightly more effective than seal 3. The overlapping radial-clearance seal 4b requires far less sealing air than the axial-clearance ones, and the difference increases as either G_c or Re_ϕ increases.

Recently, Chew (1989) has had some success in using the momentum-integral equations to estimate $C_{w,min}$ for some of the seals described above.

9.3 EFFECT OF EXTERNAL FLOW

The above tests were conducted in a quiescent environment. In a gas turbine, there is an axial flow of hot gas over the outside of the seals, and this external flow can have a significant effect on the ingress problem.

Abe, Kikuchi and Takeuchi (1979) conducted an experimental investigation of the ingress problem. They used typical turbine geometries and examined the effect of external mainstream gas flow on the sealing performance. Unlike the experiments reported above, those of Abe *et al* indicated that the amount of sealing air required to prevent ingress was independent of rotational speed but depended on the speed of the mainstream gas.

Kobayashi, Matsumato and Shizuya (1984) carried out ingress experiments using "cold" (c. 115°C) sealing air and a hot (c. 350°C) mainstream gas flow in a system with overlapping radial-clearance seals, and they observed that the temperature of the sealing air fluctuated. The fluctuation decreased and eventually decayed away as the flow-rate was increased, and they assumed that incipient ingress occurred at the onset of the fluctuations. They also measured the pressure distribution in the wheel-space, and observed the pressure-inversion effect referred to above. However, ingress determined by temperature fluctuations did not correlate with the pressure criterion, and (like the results of Abe *et al*) the values of minimum flow-rate showed relatively little increase with rotational speed over the range $0 < Re_\phi < 6.5 \times 10^6$.

There is, therefore, a contradiction between the results obtained with an external flow and those obtained without. Reasons for the difference are discussed below.

9.3.1 The datum axial-clearance seal (seal 1)

Phadke and Owen (1988b,c) conducted experiments on a rig with a rotor of 376 mm diameter. The gap ratio was kept approximately constant with $G \simeq 0.1$, and the clearance ratio was varied from $G_c = 0.005$ to 0.02. Sealing air was fed radially outwards from a central hole of 50 mm diameter in the stator, and an external flow of air was supplied axially through an annulus, of 30 mm radial height, at the periphery of the system.

For the pressure measurements, taps were arranged at 21 angular locations in the external-flow annulus. However much care was taken in aligning the annulus, there were always differences between the circumferential pressure readings. If Δp_{max} is the maximum difference between any two of the 21 pressures and $\bar{W}$ is an average axial component of velocity in the annulus, then $C_{p,max}$ is a *nondimensional pressure asymmetry* defined as

$$C_{p,max} = \frac{\Delta p_{max}}{\frac{1}{2}\rho \bar{W}^2}. \tag{9.13}$$

By creating disturbances in the annulus, it was possible to vary $C_{p,max}$, and experiments were conducted in the range $0.07 < C_{p,max} < 0.48$. As, for the rig used, it was not possible to reduce the lower limit, tests conducted with $C_{p,max} \simeq 0.07$ are referred to as "quasi-axisymmetric flows". The average radial pressure difference across the seal, $\overline{\Delta p}$, was measured by taking the arithmetic mean of the differences between each of the 21 pairs of pressure taps, which were located in the annulus and in the outer part of the stator (at $x \simeq 0.97$). Ingress was assumed to occur when the average pressure in the wheel-space was less than that in the annulus (that is, $\overline{\Delta p} < 0$). When $\overline{\Delta p}$ was zero, the resulting value of C_w was taken to be $C_{w,min}$. In addition, ingress was determined using the flow-visualization and concentration techniques described in Section 9.2.

Owing to the pressure asymmetry in the external flow, ingress was found to occur in a nonaxisymmetric manner. This could happen even when the rotor was stationary, and

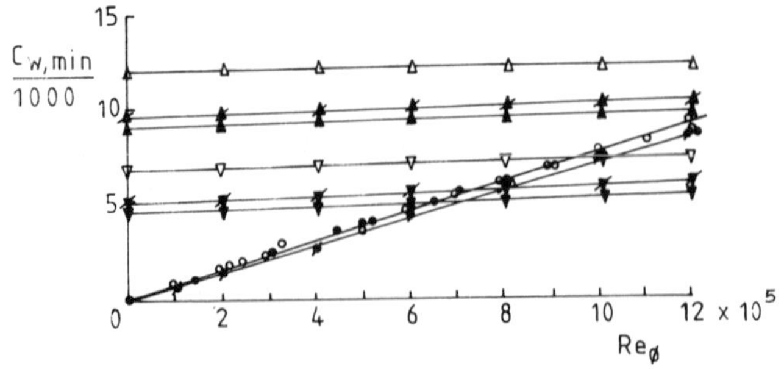

Figure 9.12 Effect of Re_w on the variation of $C_{w,min}$ with Re_ϕ for seal 1 (according to the three ingress criteria): $G_c = 0.02$ (Phadke and Owen 1988b)

$\overline{\Delta p}$ pressure criterion	○	▽	△
flow-visualization criterion	●	▼	▲
concentration criterion	○	⭑	⭑
Re_w	0	$5.9\text{x}10^5$	$1.15\text{x}10^6$

flow visualization revealed that ingested fluid moved transversely across the wheel-space from regions of high to low pressure in the annulus.

Using the three different ingress criteria, $C_{w,min}$ was determined for a range of Re_ϕ and for different values of the *external-flow Reynolds number*, Re_w, where

$$Re_w = \frac{\rho \overline{W} b}{\mu}. \tag{9.14}$$

An example is shown in Figure 9.12, for the datum axial-clearance seal (seal 1 of Figure 9.5) with a quasi-axisymmetric external flow; this illustrates the effect of Re_w on the variation of $C_{w,min}$ with Re_ϕ. Although there are some quantitative differences between the values of $C_{w,min}$ determined from the three criteria, a number of conclusions can be drawn:

(i) for $Re_w = 0$, $C_{w,min}$ increases with Re_ϕ;

(ii) for $Re_w \geq 0.59\text{x}10^6$, $C_{w,min}$ is virtually independent of Re_ϕ but increases as Re_w increases;

(iii) depending on the ratio of Re_w and Re_ϕ, external flow can either increase or decrease $C_{w,min}$.

The last of these effects can be seen clearly in Figure 9.13, which shows the effect of Re_ϕ on the variation of $C_{w,min}$ with Re_w. The results were obtained by flow visualization with quasi-axisymmetric external flow for seal 1 with $G_c = 0.02$. For small values of Re_w/Re_ϕ, there is a rotation-dominated regime in which $C_{w,min}$ increases as Re_ϕ increases. However, as the value of Re_w/Re_ϕ is increased, the effects of Re_ϕ become smaller, and at large values of Re_w/Re_ϕ there is an external-flow-dominated regime in which $C_{w,min}$ becomes independent of Re_ϕ. It is believed to be this latter regime that was observed by Abe *et al* (1979) and by Kobayashi *et al* (1984).

Considerable insight into the causes of this behaviour has been provided by Vaughan (1986), who obtained numerical solutions of the elliptic equations of motion for turbulent flow in a rotor-stator system (see Sections 6.4.1, 7.3.2). Simplified diagrams of the computed stream-

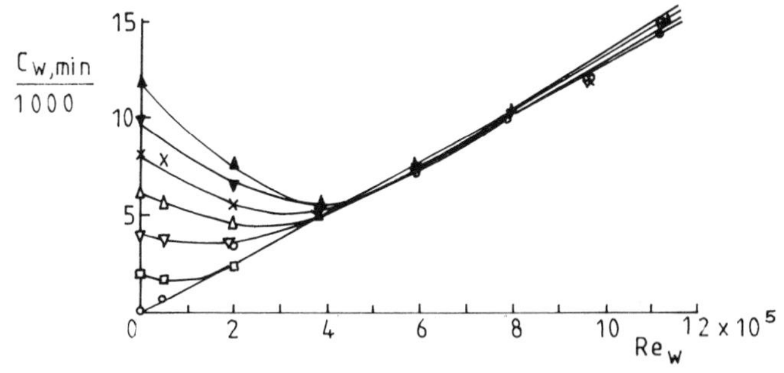

Figure 9.13 Effect of Re_ϕ on the variation of $C_{w,min}$ with Re_w for seal 1: $G_c = 0.02$ (Phadke and Owen 1988b)

Symbol	○	□	▽	△	×	▼	▲
Re_ϕ	0	2×10^5	4×10^5	6×10^5	8×10^5	10^6	1.2×10^6

lines for axisymmetric flow in the clearance are shown in Figure 9.14 for $G_c = 0.01$, $C_w = 10^3$, $Re_\phi = 8\text{x}10^5$ and three values of Re_w up to $6\text{x}10^5$. For small Re_w, external fluid is ingested into the system and flows radially inward on the edge of the shroud. For medium Re_w, a separation bubble is formed near the edge of the shroud, and the effective clearance between the shroud and the rotor is reduced. For large Re_w, the separation bubble grows in axial extent, which further restricts the effective clearance and nearly causes the axial flow to impinge on the rotor. The reduction in effective clearance causes a reduction in $C_{w,min}$ as Re_w increases: this is consistent with the experimental data at small values of Re_w/Re_ϕ shown in Figure 9.13.

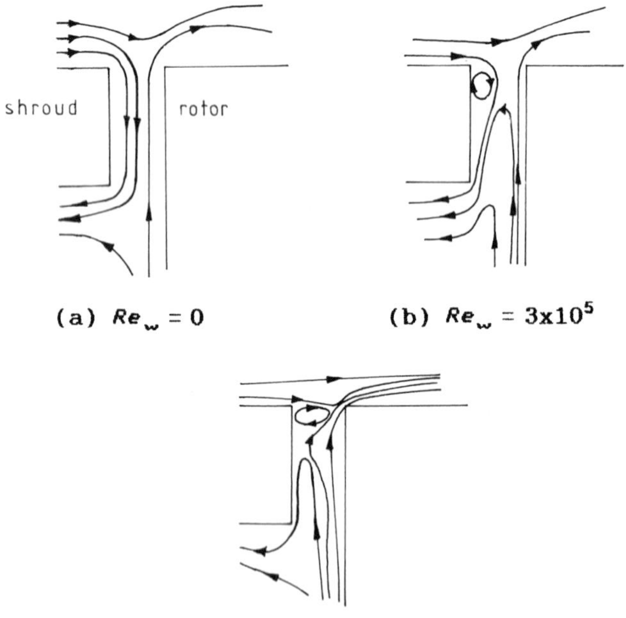

(a) $Re_w = 0$ (b) $Re_w = 3\text{x}10^5$

(c) $Re_w = 6\text{x}10^5$

Figure 9.14 Numerical prediction of the effect of Re_w on the flow in the seal clearance (Vaughan 1986)

Thus, under *axisymmetric* conditions, the effects of external flow are benign, and $C_{w,min}$ should decrease monotonically as Re_w increases. For nonaxisymmetric flow, however, the separation bubble on the stator can cause the axial flow to impinge on the rotor in some regions: this creates regions of inflow and outflow at different circumferential locations (as observed from the flow visualization discussed above). Vaughan obtained solutions of the potential-flow equations for the case where the pressure in the external flow varied periodically in the circumferential direction. Under these conditions, he showed that $C_{w,min}$ was proportional to Re_w, which is consistent with the experimental data at the larger values of Re_w/Re_ϕ shown in Figure 9.13.

It would appear, therefore, that the rotation-dominated regime, where $C_{w,min}$ decreases as Re_w increases, is caused by axisymmetric effects. Conversely, the external-flow-dominated regime, in which $C_{w,min}$ increases as Re_w increases, is caused by nonaxisymmetric effects.

In an attempt to separate the effects of the pressure asymmetry from those of Re_w, tests were carried out using six asymmetries having the circumferential pressure distributions shown in Figure 9.15 (where p is the static pressure on the outside wall of the annulus). The asymmetries, which were generated by creating blockages on a gauze screen placed upstream of the external-flow annulus, increased from $C_{p,max} \simeq 0.1$ to 0.48 for asymmetries numbers 1 to 6; for quasi-axisymmetric flow (asymmetry number 0), $C_{p,max} \simeq 0.07$.

The effect of $C_{p,max}$ on the variation of $C_{w,min}$ with Re_w is shown in Figure 9.16 for seal 1 with $G_c = 0.01$ and $Re_\phi = 0$ (in the external-flow-dominated regime). It can be seen that $C_{w,min} \propto Re_w$, and $C_{w,min}$ increases as $C_{p,max}$ increases. A simple model for the behaviour is discussed in Section 9.3.3.

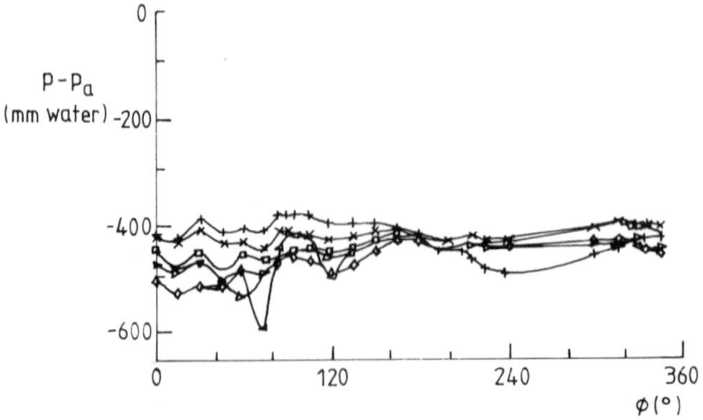

Figure 9.15 Circumferential variation of static pressure in the annulus for each of the six asymmetries: $C_w = 0$, $Re_\phi = 0$, $Re_w = 1.1 \times 10^6$ (Phadke and Owen 1988c)

Symbol	×	□	◊	▷	+	◁
Asymmetry number	1	2	3	4	5	6

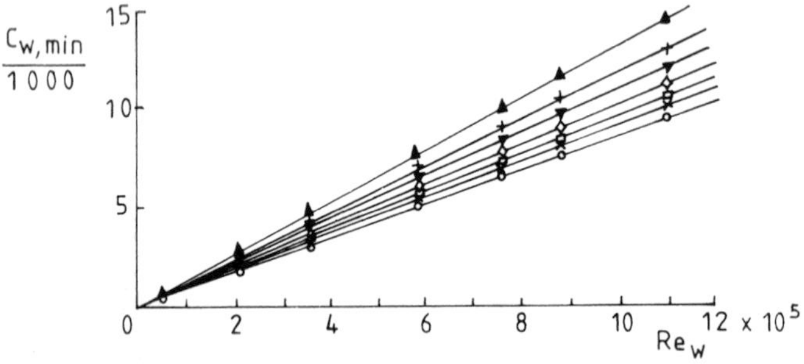

Figure 9.16 Effect of the different asymmetries on the variation of $C_{w,min}$ with Re_w for seal 1: $G_c = 0.01$, $Re_\phi = 0$ (Phadke and Owen 1988c)

Symbol	∘	×	□	◊	▼	+	▲
Asymmetry number	0	1	2	3	4	5	6
Cp,max	0.07	0.10	0.18	0.20	0.22	0.22	0.48

9.3.2 Effect of seal geometry

Tests were conducted by Phadke and Owen (1988b,c), using the external-flow rig described in Section 9.3.1, for the datum axial-clearance seal, the double-shrouded seal, the rotating axial clearance seal and the overlapping radial-clearance seal (seals 1, 2, 3 and 5 in Figure 9.5).

Figure 9.17 shows a comparison between the flow patterns observed for seals 1 and 2 with $G_c = 0.01$, $Re_\phi = 0$ and $Re_w = 0.21 \times 10^6$ (in the external-flow-dominated regime). For seal 1, ingested fluid moved transversely across the wheel-space from regions of high to low pressure in the annulus; as C_w was increased, the ingestion was reduced. For seal 2, the cavity between the inner and outer seals acted as a "ring-road" which transported most of the ingested fluid: the inner seal causes an attenuation of the circumferential pressure asymmetry generated in the external flow, and this reduces ingress in the inner wheel-space. Thus, in the external-flow-dominated regime, the double-shrouded seal has advantages that are not apparent for the tests conducted in a quiescent environment.

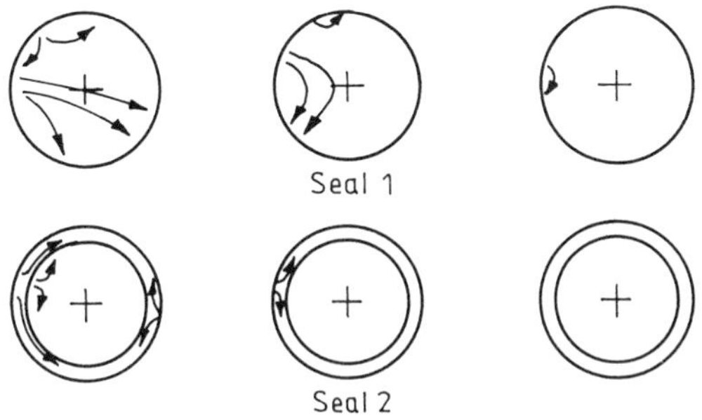

Figure 9.17 Observed flow patterns in the wheel-space: $G_c = 0.1$, $Re_\phi = 0$, $Re_w = 2.1 \times 10^5$ (Phadke and Owen 1988c)

The effect of Re_ϕ on the variation of $C_{w,min}$ with Re_w for seal 2 with $G_c = 0.01$ is shown in Figure 9.18. The values of $C_{w,min}$ were obtained for the inner and outer wheel-spaces, with asymmetry number 4 ($C_{p,max} \simeq 0.22$), using flow visualization. For $Re_w = 0$, both sets of results show similar increases of $C_{w,min}$ as Re_ϕ increases. In the external-flow-dominated regime, at large values of Re_w, the values of $C_{w,min}$ for the inner wheel-space are significantly smaller than for the outer one.

The effect of Re_ϕ on the variation of $C_{w,min}$ with Re_w for seals 1,3 and 5 with $G_c = 0.01$ is shown in Figure 9.19

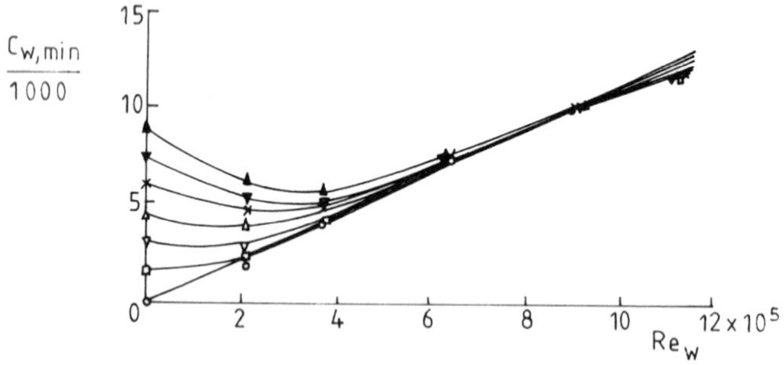

(a) Outer shroud

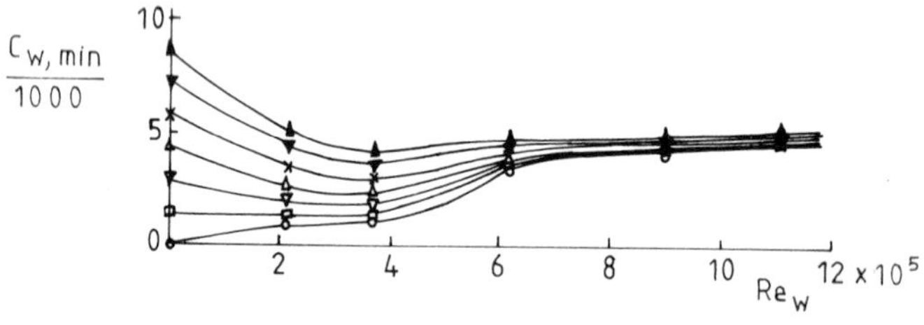

(b) Inner shroud

Figure 9.18 Effect of Re_ϕ on the variation of $C_{w,min}$ with Re_w for seal 2 with asymmetry number 4: $G_c = 0.01$ (Phadke and Owen 1988c)

Symbol	○	□	▽	△	×	▼	▲
Re_ϕ	0	$2x10^5$	$4x10^5$	$6x10^5$	$8x10^5$	10^6	$1.2x10^6$

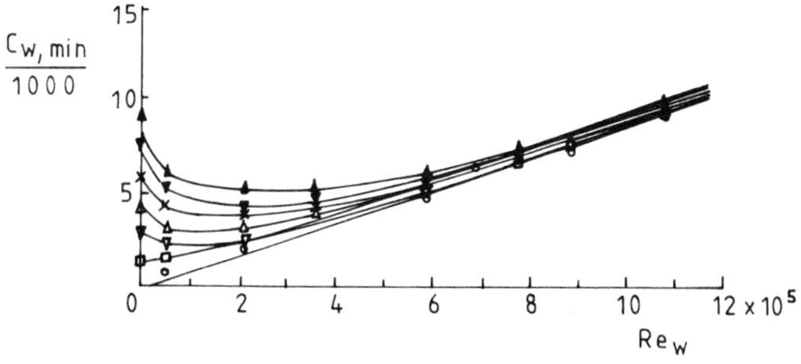

(a) Seal 1

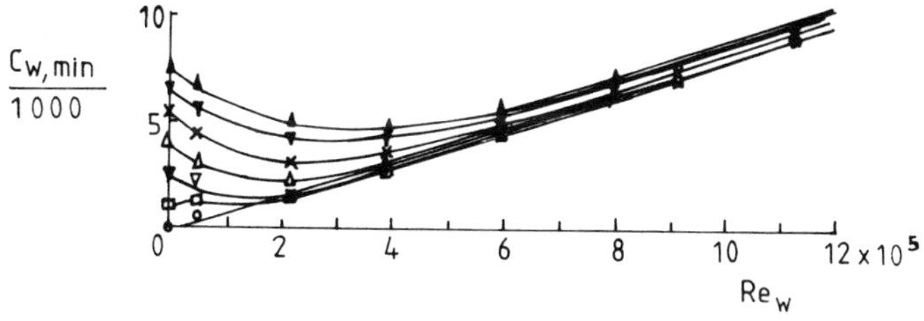

(b) Seal 3

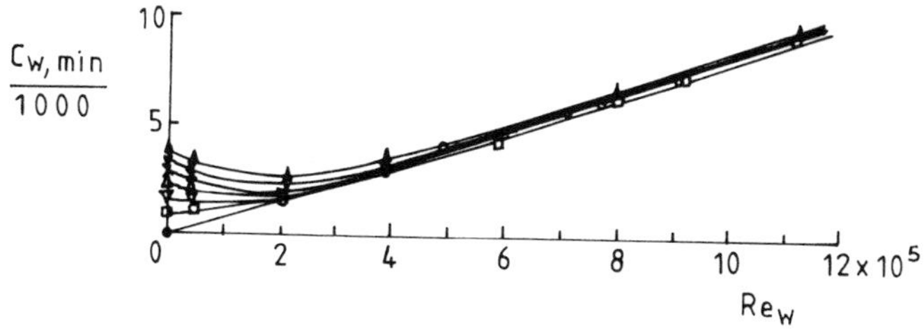

(c) Seal 5

Figure 9.19 Comparison of the effects of Re_ϕ on the variation of $C_{w,min}$ with Re_w for seals 1, 3 and 5 with asymmetry number 0: $G_c = 0.01$ (Phadke and Owen 1988c)

Symbol	∘	□	▽	△	×	▼	▲
Re_ϕ	0	2×10^5	4×10^5	6×10^5	8×10^5	10^6	1.2×10^6

for asymmetry number 0 ($C_{p,max} \simeq 0.07$). The order-of-merit of the seals is 5, 3, 1 (as it was in the quiescent environment), but the advantages of the radial-clearance seal 5 appears to be smaller in the external-flow-dominated regime than in the rotation-dominated one. Additional tests using concentration measurements suggested that the amount of fluid ingested into the wheel-space is likely to be less for seal 5 than for the axial-clearance seals.

From the above results, it would appear that a double-shrouded radial-clearance seal would have advantages to offer in both the rotation-dominated and the external-flow-dominated regimes. (Haaser, Jack and McGreehan 1988 recommended a similar type of seal for gas-turbine applications.)

Graber, Daniels and Johnson (1987) conducted tests to determine the effectiveness of four different seal configurations at $Re_\phi = 2.5\text{x}10^6$ and $5\text{x}10^6$. They used carbon dioxide as a tracer gas to estimate the amount of external air that was ingested into the "downstream wheel-space" (that is, the external air flowed axially from the rotor towards the stator, in contrast to the tests described above). It is difficult to make direct comparisons with the data of these authors, but the amount of air necessary to seal their system appeared to be similar in magnitude to that obtained from the correlations given above.

9.3.3 Simple model for the calculation of $C_{w,min}$

In the external-flow-dominated regime, it is assumed that the sealing mass flow-rate, $\dot{m}$, is related to the circumferentially-averaged pressure drop across the shroud, Δp_s, by

$$\dot{m} = C_d A \, (2\rho \Delta p_s)^{\frac{1}{2}}, \qquad (9.15)$$

where $A = 2\pi b s_c$ is the area of the seal opening and C_d is a discharge coefficient. Equation (9.15) can be written in terms of nondimensional groups as

$$C_w = 2\sqrt{2}\,\pi C_d G_c \left(\frac{\rho \Delta p_s b^2}{\mu^2}\right)^{\frac{1}{2}}. \qquad (9.16)$$

Ingress is related to the amplitude of the circumferential pressure differences in the external flow, and it is postulated that $C_w = C_{w,min}$ when $\Delta p_s = A\,\Delta p_{max}$, where Δp_{max} is the maximum circumferential pressure difference and A is an empirical constant. Equation (9.16) then gives

$$C_{w,min} = 2\pi K\, G_c\, P_{max}^{\frac{1}{2}}, \qquad (9.17)$$

where

$$K = (2A)^{\frac{1}{2}} C_d, \qquad P_{max} = \frac{\rho \Delta p_{max} b^2}{\mu^2} = \frac{1}{2} C_{p,max} Re_w^2. \qquad (9.18)$$

Figure 9.20 shows a comparison between equation (9.17) and the data measured by Phadke and Owen (1988c) for seals 1, 2, 3 and 5 (see Figure 9.5). The experimental results were obtained using flow-visualization for $G_c = 0.005$, 0.01 and 0.02, $Re_\phi = 0$ and $Re_w = 1.1\text{x}10^6$; for seal 2, only the results for the outer wheel-space are included. Despite the scatter, the results are correlated reasonably well using a value of $K = 0.6$ in equation (9.17).

It should be emphasized that this simple model takes no account of seal geometry, of the circumferential distribution of pressure nor of swirl in the external flow. However, used in conjunction with the results for the rotation-dominated regime presented in Section 9.2, it should provide an estimate for $C_{w,min}$.

9.4 EFFECT OF BLADE-COOLING FLOW

As discussed in Section 8.3.2 in an engine there is often a radial outflow of air over the turbine disc and an axial flow of air from nozzles in the outer part of the stator to holes in the rotor. The radial outflow is used to cool the disc and to prevent ingress, and the axial flow is

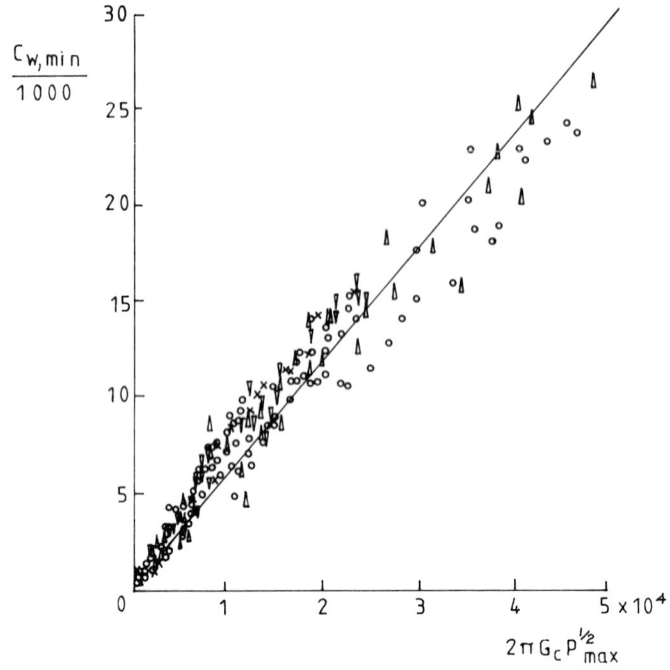

Figure 9.20 Variation of $C_{w,min}$ with $2\pi G_c P^{1/2}_{max}$ (Phadke and Owen 1988c)

———	Equation (9.17): K = 0.6			
Symbol	∘	×	∆	∇
Seal	1	2	3	5

used to provide cooling air for the turbine blades attached to the periphery of the disc. The two flows can interact, and the amount of blade-coolant used can affect the amount of air required to prevent ingress. This problem was studied by El-Oun (1986) and by El-Oun, Neller and Turner (1988) using the apparatus described in Section 8.3.2 and shown in Figure 8.11.

Computations of the laminar axisymmetric case were conducted by El-Oun using an elliptic solver with a nonuniform 59x65 grid. The nozzles in the stator and the holes in the rotor, at $x = 0.9$, were modelled using annular slots; whilst this cannot represent the three-dimensional flow that occurs inside the engine, it does give some insight into the flow structure. Figure 9.21 shows the computed streamlines for $G = 0.1$, $Re_\phi = 5\times10^4$, $C_{w,d} = C_{w,s} = 0$ and $C_{w,b} = C_{w,p} = 100$ to 1100; the subscripts b, d, p and s are used to refer to the blade-cooling, disc-cooling, pre-swirl and sealing flows respectively.

Referring to Figure 9.21, it can be seen that for $C_{w,p} = 100$ all the pre-swirl air flows radially inward on the stator and outward on the rotor before it enters the blade-cooling slot in the rotor. For $C_{w,p} = 1100$, the pre-swirl air flows axially across the wheel-space from the

---

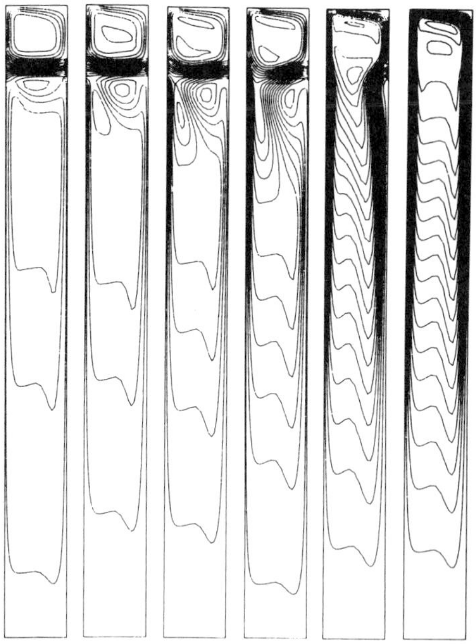

$C_{w,p} =$ 1100 900 700 500 300 100

Figure 9.21 Computed streamlines for pre-swirl feed system: $G = 0.1$, $Re_\phi = 5\times10^4$ (El-Oun 1986)

rotor to the stator, with separate recirculation zones on either side of the annular jet. According to the approximation given by equation (3.64), the nondimensional flow-rate, $C_{w,0}$, entrained in the boundary layer of an enclosed laminar rotor-stator system (for $x = 0.9$, $Re_\varphi = 5x10^4$ and $\beta^* = 0.382$) is given by $C_{w,0} \simeq 350$. It is interesting to observe from Figure 9.21 that, for $C_{w,p} \leq 300$, the pre-swirl flow fails to penetrate directly across the core between the two boundary layers. Thus, as for the impingement heating discussed in Section 8.3.1, the ratio of the impinging flow-rate to the entrained flow-rate is likely to be an important parameter.

In the rig, experiments were conducted with and without inner seals fitted. The clearance ratio of the outer mitred seals was maintained at either $G_c = 0.005$ or $G_c = 0.0072$, and the radial-clearance ratio of the inner seal, when fitted, was $G_c = 0.0033$. Ingress was determined using criteria based on pressure and concentration measurements as well as on flow visualization, as described in Section 9.3. In addition, for some tests conducted without the inner seals, a temperature criterion was used. This was based on the temperature difference between the air outside the system, which was heated by hot-air blowers, and the unheated air immediately inside. Values of $C_{w,min}$ obtained using two or more of the four criteria were found to be in satisfactory agreement.

Figure 9.22 shows the variation of $C_{w,min}$ with Re_φ for the case where $G_c = 0.0072$, $C_{w,b} = 0$ and the inner seal is removed. The pressure and flow-visualization criteria both show that $C_{w,min}$ is not significantly affected by the position at which the sealing flow is introduced. For these conditions (where $C_{w,min}$ is smaller than the entrained value, $C_{w,0}$) the pre-swirl air probably flows radially inward on the stator, as was observed for laminar flow in Figure 9.21 when $C_{w,p}/C_{w,0} < 1$.

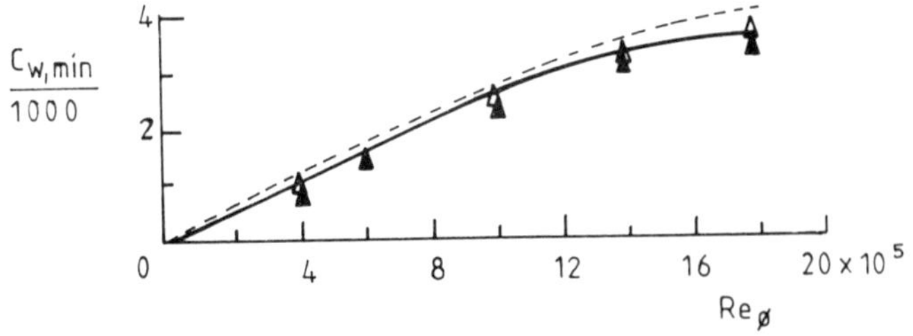

Figure 9.22 Variation of $C_{w,min}$ with Re_ϕ for the pre-swirl rig: $G_c = 0.0072$, $C_{w,b} = 0$ (El-Oun 1986)

———	pressure criterion, $C_{w,d} = 0$
▲	flow visualization, $C_{w,d} = 0$
- - - - -	pressure criterion, $C_{w,p} = 0$
Δ	flow visualization, $C_{w,p} = 0$

Figure 9.23 shows the variation of $C_{w,min}$ with $C_{w,b}$ (obtained by the pressure criterion) for $G_c = 0.0072$, $C_{w,d} = C_{w,min}$ and $C_{w,p} = C_{w,b}$ for the case where no inner seals are fitted. It can be seen that, for the smaller values of $C_{w,b}$, $C_{w,min}$ tends to decrease as $C_{w,b}$ increases; for larger values of $C_{w,b}$, $C_{w,min}$ is, for $Re_\phi > 0$, only weakly dependent on Re_ϕ. It can also be seen that ingress can occur when $Re_\phi = 0$. This behaviour is similar to that observed in the rotation-dominated and external-flow-dominated regimes discussed in Section 9.2. With the inner seal fitted, it was found that the value of $C_{w,min}$ for the inner wheel-space was only weakly affected by the value of $C_{w,b}$: a similar result was observed for the double-shrouded system with an external flow (as discussed in Section 9.3.2). The axial flow from the pre-swirl nozzles appears to act like a nonaxisymmetric external flow.

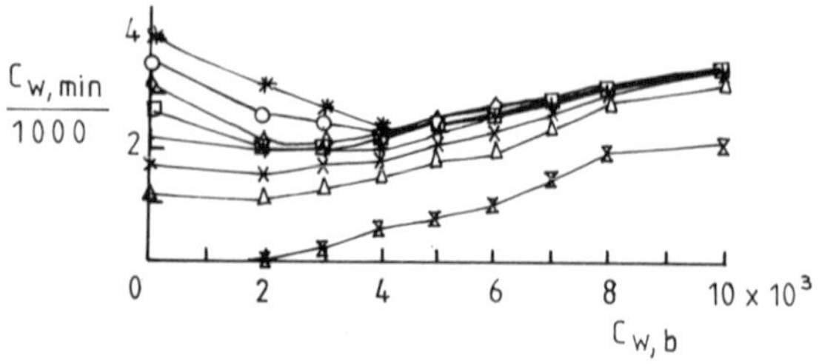

Figure 9.23 Effect of Re_ϕ on the variation of $C_{w,min}$ with $C_{w,b}$ for the pre-swirl rig: $G_c = 0.0072$, $C_{w,d} = C_{w,min}$, $C_{w,p} = C_{w,b}$ (El-Oun *et al* 1988)

Symbol	⧗	△	×	+	□	◊	○	✱
Re_ϕ	0	$4x10^5$	$6x10^5$	$8x10^5$	10^6	$1.2x10^6$	$1.4x10^6$	$1.8x10^6$

9.5 CONCLUDING REMARKS

Although experiments provide insight into the ingress problem and produce data of use to the design engineer, the computation of such flows presents serious difficulties. A complete solution of the ingress problem (and of many other problems in rotating-disc systems) requires the use of three-dimensional turbulent elliptic-solvers, and at high Reynolds numbers such solutions have been, so far, prohibitively expensive in computing time.

In the Thermo-Fluid Mechanics Research Centre, recent experience with multigrid techniques (see Section 7.3.2) suggests that accurate and economical solutions of three-dimensional rotating flows may soon be practicable at conditions representative of those found in modern gas turbines. If this expectation is realized, it will not be long before some of the more intractable problems are solved: with luck, the difficult old problems will be replaced by new ones that are challenging enough to satisfy the most ambitious research worker!

APPENDIX A

Numerical Solution of Generalized Von Kármán Equations

A.1 The basic equations

The functions $F(\zeta)$, $G(\zeta)$ and $H(\zeta)$ are functions of the independent variable ζ; they satisfy the equations

$$F'' = F^2 - G^2 + F'H + \beta^2 \tag{A.01}$$

$$G'' = 2FG + G'H \tag{A.02}$$

$$H' = -2F \tag{A.03}$$

(where primes denote differentiations with respect to ζ), and are subject to the boundary conditions

$$F(0) = 0, \quad G(0) = G_0, \quad H(0) = 0 \tag{A.04}$$

$$F(\zeta) \to 0, \quad G(\zeta) \to G_\infty \quad \text{as } \zeta \to \infty. \tag{A.05}$$

The constants G_0, G_∞ and β are assumed to be given.

Equations of this form occur in three places in the main text:

(i) von Kármán's equations for the free disc (equations (3.05) to (3.10)), for which $\beta = 0$, $G_0 = 1$ and $G_\infty = 0$;

(ii) the equations for a rotating disc in a rotating fluid (see Section 3.2.1), for different values of β and with $G_0 = 1$, $G_\infty = \beta$;

(iii) the equations for a stationary disc in a rotating fluid (see Section 3.2.2), where $\beta = 1$, $G_0 = 0$ and $G_\infty = 1$.

It is of interest to determine the dependence of $F(\zeta)$, $G(\zeta)$ and $H(\zeta)$ on ζ; but it is of particular importance to the engineer to determine $F'(0)$, $G'(0)$ and H_∞ (the limiting value of $H(\zeta)$ as $\zeta \to \infty$).

A.2 The initial-value problem

If $F'(0)$ and $G'(0)$ are known, as well as the conditions (A.04), the equations (A.01) to (A.03) can be solved as an initial value problem. The method described here (used first by Rogers and Lance 1960) is an iterative method in which approximate values of $F'(0)$and $G'(0)$ are assumed; a systematic trial-and-error method is then used to modify these values until the boundary conditions (A.05) are satisfied for a sufficiently large value of ζ.

An alternative condition for $\zeta \to \infty$ involves the functions (see Zandbergen and Dijkstra 1987)

$$\phi(\zeta) = F'^2(\zeta) + G'^2(\zeta), \qquad \psi(\zeta) = FF'' + G'^2. \qquad \text{(A.06)}$$

It can be shown that $\phi'(\zeta)$ cannot change sign more than once, and that $\psi(\zeta)$ cannot change sign more than twice. Since, for the required solution, $\phi(\zeta) \to 0$ and $\psi(\zeta) \to 0$ as $\zeta \to \infty$, the approach must be monotonic. In practice it is found that $\phi(\zeta)$ and $|\psi(\zeta)|$ decrease as ζ increases up to $\zeta = \zeta_0$ (say); then, for $\zeta > \zeta_0$, one or other (or both) begins to increase. It is clear that a second unwanted solution is present in the numerical solution and there is no point in proceeding further. The intrusion of the unwanted solution is due either to incorrect values of $F'(0)$ and $G'(0)$ or to the accumulation of rounding errors. In practice, an attempt is made to eliminate the first possible error so as to make ζ_0 as large as possible. (This monotonic approach to zero is easier to monitor than the approach of $F'(\zeta)$ or $G'(\zeta)$ which, except for the case of the free disc, is oscillatory.)

The approach of $H(\zeta)$ to its limiting value, H_∞, is slower than that of $\phi(\zeta)$ and $\psi(\zeta)$; it is, therefore, necessary to use the asymptotic solution for large ζ, which is discussed in Section 3.2.1. Using equation (3.52) together with equation (A.03), it is not difficult to show that

$$H(\zeta) \sim H_\infty - \frac{2}{\lambda_1^2 + \lambda_2^2}[\lambda_1 F(\zeta) + \lambda_2 (G - \beta)] \quad \text{as } \zeta \to \infty, \tag{A.07}$$

where

$$\lambda_1 = \tfrac{1}{2}\left[H_\infty - \left(\tfrac{1}{2}[H_\infty^2 + (H_\infty^4 + 64\beta^2)^{\frac{1}{2}}]\right)^{\frac{1}{2}}\right], \tag{A.08}$$

$$\lambda_2 = \beta/(\lambda_1 - \tfrac{1}{2}H_\infty). \tag{A.09}$$

These three equations can be used to find an estimate of H_∞ using the computed values of F, G and H for each value of ζ. When β is small, equation (A.07) is adequate to give an accurate value of H_∞ (determined by the fact that successive values of the estimate of H_∞ are unchanged). As β increases, it is necessary to use a second-order asymptotic solution; in this case equation (A.07) must be replaced by

$$H = H_\infty - \frac{2}{\lambda_1^2 + \lambda_2^2}[\lambda_1 F + \lambda_2 (G - \beta)] - \frac{\lambda_2}{\beta(4\lambda_1^2 + \lambda_2^2)}[F^2 + (G - \beta)^2]. \tag{A.10}$$

(The derivation of this equation is not given here.)

A.3 The system of first-order equations

At each stage of the iterative technique to determine $F'(0)$ and $G'(0)$ accurately (and, hence, H_∞), it is convenient to write

$$F'(0) \simeq a, \quad G'(0) \simeq b; \tag{A.11}$$

here a and b are constants (which have no connection with disc radii) that must be specified. Initially they are chosen by using, for example, the results of Rogers and

Lance (1960) which, though not as accurate as those authors claimed, are good approximations. After the first computation, the values of a and b are adjusted in a systematic way (usually one decimal place at a time) so that ζ_0 increases.

The equations are put into a standard form, suitable for numerical computation, by defining the functions

$$y_1 = F,\ y_2 = F',\ y_3 = G,\ y_4 = G',\ y_5 = H. \tag{A.12}$$

It is then easy to demonstrate that equations (A.01) to (A.03), together with the initial conditions (A.04) and $F'(0) = a$, $G'(0) = b$, become

$$y_1' = y_2, \quad y_1(0) = 0, \tag{A.13}$$

$$y_2' = y_1^2 - y_3^2 + y_2 y_5 + \beta^2, \quad y_2(0) = a, \tag{A.14}$$

$$y_3' = y_4, \quad y_3(0) = G_0, \tag{A.15}$$

$$y_4' = 2y_1 y_3 + y_4 y_5, \quad y_4(0) = b, \tag{A.16}$$

$$y_5' = -2y_1, \quad y_5(0) = 0. \tag{A.17}$$

Since the constants in this system of equations are all specified (at each stage of the iteration), the equations can be solved using a suitable standard computer-library routine. The simplest such routine in a variable-step Runge-Kutta one, such as that used by Rogers and Lance; this is slow and difficulties may arise if there is stiffness in the system over part of the range. Most of the computations described here have been carried out using a variable-step Gear technique which is faster and allows for stiffness; for small values of β this has produced excellent results, but (for reasons not completely understood, but possibly related to the fact that the technique treats the system as stiff everywhere, even if stiffness occurs only in part of the range of integration) as β increased, convergence to the required accuracy was impossible to achieve. The Shampine-Gordon (1975) technique has proved more successful for the Bödewadt problem (a stationary disc in a rotating fluid); it has not yet been tried for the case of a rotating disc in a rotating fluid.

It has been found that, using the Shampine-Gordon method on a VAX 8530 computer with a typical precision of 16 significant figures, integration cannot proceed further than $\zeta \simeq 13$ for the Bödewadt problem. For the free disc, a and b must be computed to an accuracy of at least 10 significant figures to ensure that the error in $|H - H_\infty|$ is less than 10^{-7} at "large" values of ζ. The Gear technique, used on the VAX 8530, showed evidence of signficant accumulated rounding errors for $\zeta \gtrsim 20$; increasing the accuracy of a and b gave no improvement in either the value of $|H - H_\infty|$ or the value of ζ at which rounding errors became important.

A.4 Benton's method for the free disc

For a disc (either rotating or stationary) in a rotating fluid, there is no alternative but to use the equations in the form given in equations (A.01) to (A.05). For the free disc, however, the method described by Benton (1964) transforms the infinite ζ-range to a finite ξ-range (see equations (3.14) to (3.20)). The integration can be carried out in one step from one end of the range to the other, and problems of accumulated rounding errors do not arise. The method described above can be used, except that the first and third of conditions (3.20) are used instead of inspection of the functions ϕ and ψ.

APPENDIX B

Numerical Solution of Energy Equation

B.1 The function $\theta(\zeta)$ for the free disc

The function $\theta(\zeta)$ (see equation (5.44)) satisfies the equation

$$\theta'' - Pr\,(H\theta' + nF\theta) = 0, \tag{B.01}$$

where Pr and n are given constants and $F(\zeta)$ and $H(\zeta)$ are two of the functions discussed in Appendix A. The boundary conditions (see equation (5.37)) are

$$\theta(0) = 1, \quad \theta(\zeta) \to 0 \text{ as } \zeta \to \infty. \tag{B.02}$$

It is not difficult to show that the general solution of equation (B.01) can be written in the form

$$\theta(\zeta) = A\,\theta_1(\zeta) + B\,\theta_2(\zeta), \tag{B.03}$$

where

$$\theta_1'' - Pr\,(H\theta_1' + nF\theta_1) = 0, \quad \theta_1(0) = 1, \; \theta_1'(0) = 0, \tag{B.04}$$

$$\theta_2'' - Pr\,(H\theta_2' + nF\theta_2) = 0, \quad \theta_2(0) = 0, \; \theta_2'(0) = 1, \tag{B.05}$$

and A, B are arbitrary constants. When the boundary conditions (B.02) are applied to the solution (B.03), it is easy to see that

$$A = 1, \quad B = \lim_{\zeta\to\infty}\left[\frac{\theta_1(\zeta)}{\theta_2(\zeta)}\right]. \tag{B.06}$$

It is convenient to write $y_6(\zeta)=\theta_1(\zeta)$, $y_7(\zeta)=\theta_1'(\zeta)$, $y_8(\zeta)=\theta_2(\zeta)$ and $y_9(\zeta)=\theta_2'(\zeta)$. Then equations (B.04) and (B.05) can be written in the form

$$y_6' = y_7, \quad y_6(0)=1, \tag{B.07}$$

$$y_7' = Pr\,(y_5y_7 + ny_1y_6), \quad y_7(0)=0, \tag{B.08}$$

$$y_8' = y_9, \quad y_8(0)=0, \tag{B.09}$$

$$y_9' = Pr\,(y_5y_9 + ny_1y_8), \quad y_9(0)=1, \tag{B.10}$$

where $y_1(\zeta)=F(\zeta)$ and $y_5(\zeta)=H(\zeta)$, as in Appendix A. Equations (B.07) to (B.10), together with equations (A.13) to (A.17) form a well-posed system of 9 first order equations with appropriate initial conditions at $\zeta=0$. The constants a and b in equations (A.14) and (A.16) are, of course, replaced by the values of $F'(0)$ and $G'(0)$ respectively, which have already been computed; very accurate values are required to ensure that the computed value of $H(\zeta)$ is accurate (this is particularly important for small values of Pr). Suitable values of a and b are

$$a = 0.510232618867, \quad b = -0.615922014399, \tag{B.11}$$

which are correct to 12 significant figures.

The value of B, which involves a limit as $\zeta\to\infty$, can be difficult to determine for small values of Pr when the thermal boundary layer is very thick; accumulated rounding errors become significant in the computation. The problem can be overcome by considering the asymptotic behaviour of $\theta_1(\zeta)$ and $\theta_2(\zeta)$ as $\zeta\to\infty$: it is assumed that, for sufficiently large ζ, $F=0$ and $H=H_\infty$, whose value has been determined in Appendix A. If the asymptotic solution is assumed to be valid for $\zeta\geq\zeta_0$, it is easy to show that

$$\lim_{\zeta\to\infty}[\theta_1(\zeta)] = \tilde{\theta}_1(\zeta_0), \quad \lim_{\zeta\to\infty}[\theta_2(\zeta)] = \tilde{\theta}_2(\zeta_0), \tag{B.12}$$

where

$$\tilde{\theta}_1(\zeta) = \theta_1(\zeta) - \frac{1}{\mathcal{P}r H_\infty}\theta_1''(\zeta), \quad \tilde{\theta}_2(\zeta) = \theta_2(\zeta) - \frac{1}{\mathcal{P}r H_\infty}\theta_2''(\zeta). \quad (B.13)$$

This limit is attained for relatively small values of ζ: for $n = 2$ and values of $\mathcal{P}r$ between 10^{-6} and 10^9, it has been found (using a VAX 8530 computer with a typical accuracy of 16 significant figures) that the limit is reached, correct to six significant figures, before $\zeta = 20$.

The form of the function $\theta(\zeta)$ is of academic interest; the value of $\theta'(0)$, which is required for the prediction of heat transfer, is of practical interest. Differentiation of equation (B.03) shows that

$$\theta'(0) = B = -\frac{\tilde{\theta}_1(\zeta_0)}{\tilde{\theta}_2(\zeta_0)} = -\lim_{\zeta\to\infty}\left[\frac{\tilde{\theta}_1(\zeta)}{\tilde{\theta}_2(\zeta)}\right], \quad (B.14)$$

where $\tilde{\theta}_1(\zeta)$ and $\tilde{\theta}_2(\zeta)$ are given by equation (B.13).

B.2 The functions $\alpha(\zeta)$, $\beta(\zeta)$, $\gamma(\zeta)$, $\epsilon(\zeta)$ and $\eta(\zeta)$

The method used for determining the function $\theta(\zeta)$ can be extended to solve equations (5.11) to (5.17) for the functions $\alpha(\zeta)$, $\beta(\zeta)$, $\gamma(\zeta)$, $\epsilon(\zeta)$ and $\eta(\zeta)$.

It should be noted first that the complementary functions for $\beta(\zeta)$ and $\gamma(\zeta)$ satisfy the equation for $\alpha(\zeta)$, and that for $\eta(\zeta)$ the equation for $\epsilon(\zeta)$. Further, the solution of the equation for $\alpha(\zeta)$ having the initial conditions $\alpha(0) = 1$ and $\alpha'(0) = 0$ is $\alpha(\zeta) = \alpha_1(\zeta) = 1$.

It is then easy to show that

$$\alpha(\zeta) = 1 + B_\alpha \alpha_2(\zeta), \quad (B.15)$$

$$\epsilon(\zeta) = \epsilon_1(\zeta) + B_\epsilon \epsilon_2(\zeta), \quad (B.16)$$

$$\eta(\zeta) = \eta_1(\zeta) + B_\eta \epsilon_2(\zeta), \quad (B.17)$$

where

$$\alpha_2^{\zeta\prime} - Pr\, H\alpha_2^{\zeta} = 0, \quad \alpha_2(0) = 0, \; \alpha_2^{\zeta}(0) = 1, \tag{B.18}$$

$$\epsilon_1^{\zeta\prime} - Pr\,(H\epsilon_1^{\zeta} + 2F\epsilon_1) = 0, \quad \epsilon_1(0) = 1, \; \epsilon_1^{\zeta}(0) = 0 \tag{B.19}$$

$$\epsilon_2^{\zeta\prime} - Pr\,(H\epsilon_2^{\zeta} + 2F\epsilon_2) = 0, \quad \epsilon_2(0) = 0, \; \epsilon_2^{\zeta}(0) = 1, \tag{B.20}$$

$$\eta_1^{\zeta\prime} - Pr\,(H\eta_1^{\zeta} + 2F\eta_1) = -Pr\,(F'^2 + G'^2), \quad \eta_1(0) = 0, \; \eta_1^{\zeta}(0) = 0 \tag{B.21}$$

and

$$B_\alpha = -\lim_{\zeta\to\infty}\left[\frac{1}{\tilde{\alpha}_2(\zeta)}\right], \quad B_\epsilon = -\lim_{\zeta\to\infty}\left[\frac{\tilde{\epsilon}_1}{\tilde{\epsilon}_2(\zeta)}\right], \quad B_\eta = -\lim_{\zeta\to\infty}\left[\frac{\tilde{\eta}_1}{\tilde{\epsilon}_2(\zeta)}\right]. \tag{B.22}$$

In these equations

$$\tilde{\alpha}_2(\zeta) = \alpha_2(\zeta) - \frac{1}{Pr\,H_\infty}\alpha_2^{\zeta}(\zeta), \tag{B.23}$$

$$\tilde{\epsilon}_1(\zeta) = \epsilon_1(\zeta) - \frac{1}{Pr\,H_\infty}\epsilon_1^{\zeta}(\zeta), \tag{B.24}$$

$$\tilde{\epsilon}_2(\zeta) = \epsilon_2(\zeta) - \frac{1}{Pr\,H_\infty}\epsilon_2^{\zeta}(\zeta), \tag{B.25}$$

$$\tilde{\eta}_1(\zeta) = \eta_1(\zeta) - \frac{1}{Pr\,H_\infty}\eta_1^{\zeta}(\zeta), \tag{B.26}$$

as for the corresponding functions $\tilde{\theta}_1(\zeta)$ and $\tilde{\theta}_2(\zeta)$ discussed in Section B.1.

It can also be shown that the functions $\beta(\zeta)$ and $\gamma(\zeta)$ are of the form

$$\beta(\zeta) = \beta_1(\zeta) + B_\eta\gamma_2(\zeta) + B_\beta\alpha_2(\zeta) \tag{B.27}$$

$$\gamma(\zeta) = \gamma_1(\zeta) + B_\epsilon\gamma_2(\zeta) + B_\gamma\alpha_2(\zeta) \tag{B.28}$$

where

$$\beta_1^{\zeta\prime} - Pr\,H\beta_1^{\zeta} = -4\eta_1 - 12\,Pr F^2, \tag{B.29}$$

$$\gamma_1^{\zeta\prime} - Pr\,H\gamma_1^{\zeta} = -4\epsilon_1, \tag{B.30}$$

$$\gamma_2^{\zeta\prime} - Pr\,H\gamma_2^{\zeta} = -4\epsilon_2, \tag{B.31}$$

and

$$B_\beta = -\lim_{\zeta\to\infty}\left[\frac{\tilde{\beta}_1(\zeta) + \tilde{B}_\eta\gamma_2(\zeta)}{\tilde{\alpha}_2(\zeta)}\right], \tag{B.32}$$

$$B_\gamma = -\lim_{\zeta\to\infty}\left[\frac{\tilde{\gamma}_1(\zeta) + B_\epsilon\tilde{\gamma}_2(\zeta)}{\tilde{\alpha}_2(\zeta)}\right]. \tag{B.33}$$

The functions $\tilde{\beta}_1(\zeta)$, $\tilde{\gamma}_1(\zeta)$ and $\tilde{\gamma}_2(\zeta)$ are given by

$$\tilde{\beta}_1(\zeta) = \beta_1(\zeta) - \frac{1}{Pr H_\infty}\beta_1^\zeta - \frac{1}{Pr^3 H_\infty^3}\eta_1(\zeta), \qquad (B.34)$$

$$\tilde{\gamma}_1(\zeta) = \gamma_1(\zeta) - \frac{1}{Pr H_\infty}\gamma_1^\zeta - \frac{1}{Pr^3 H_\infty^3}\epsilon_1(\zeta), \qquad (B.35)$$

$$\tilde{\gamma}_2(\zeta) = \gamma_2(\zeta) - \frac{1}{Pr H_\infty}\gamma_2^\zeta - \frac{1}{Pr^3 H_\infty^3}\epsilon_2(\zeta) \qquad (B.36)$$

These expressions are obtained by taking account of the asymptotic forms of $\epsilon_1(\zeta)$, $\epsilon_2(\zeta)$ and $\eta_1(\zeta)$ as well as those of $F(\zeta)$ and $H(\zeta)$. (It is of interest to note that if the asymptotic form given in equation (B.34) is not used, it is necessary to integrate as far as $\zeta = 87$ to obtain an accuracy of three significant figures in B_β when $Pr = 0.1$; for $Pr = 0.01$, integration as far as $\zeta = 300$ gives a value of B_β accurate to only one significant figure.)

The heat transfer can be computed using the values

$$\alpha'(0) = B_\alpha, \ \beta'(0) = B_\beta, \ \gamma'(0) = B_\gamma, \qquad \epsilon'(0) = B_\epsilon, \ \eta'(0) = B_\eta. \qquad (B.37)$$

APPENDIX C

Energy Equation for Small and Large Values of Prandtl Number

C.1 Introduction

For very large and very small values of Pr the thicknesses of the thermal and the velocity boundary layers on a disc are of different orders of magnitude; the equations of Appendix B can then be very much simplified. The asymptotic solutions presented here give values of $\alpha'(0)$, $\beta'(0)$, $\gamma'(0)$, $\epsilon'(0)$ and $\eta'(0)$ which are in error by about 1% for $Pr = 0.001$ and for $Pr = 10^6$. (The asymptotic value of $\beta'(0)$ as $Pr \to \infty$ is an exception, and is not predicted by the method described below.) The asymptotic solutions are of little practical use, but can be compared with the full solution for $Pr < 10^{-3}$ or $Pr > 10^6$ and serve as a useful check on the accuracy of the computing technique used. (Asymptotic forms for $\alpha(\zeta)$ were given by Millsaps and Pohlhausen 1952.)

C.2 Small values of the Prandtl number

In this case, the thermal boundary layer is very much thicker than the velocity one; throughout most of the thermal layer, therefore, the axial component of velocity is very close to the asymptotic value, H_{∞}, as $\zeta \to \infty$. It follows that, to a good approximation (which improves as Pr decreases),

$$\int_0^\infty [H(\zeta) - H_\infty] f(\zeta)\, d\zeta \to 0 \quad \text{as } Pr \to 0 \qquad \text{(C.01)}$$

for any finite function $f(\zeta)$. For the free-disc case, in which $H(\zeta)$ is a monotonic function of ζ, the magnitude of the indefinite integral of $[H(\zeta) - H_\infty]$ cannot be greater than the integral on the left-hand side of equation (C.01); it is, therefore, permissible to neglect this integral whenever necessary.

Before discussing the equations themselves, it is convenient to define two indefinite integrals and their limits, I_∞ and J_∞ as $\zeta \to \infty$:

$$I(\zeta) = \int_0^\zeta [F'^2(\xi) + G'^2(\xi)] d\xi, \qquad I(\zeta) \to I_\infty \quad \text{as } \zeta \to \infty \qquad \text{(C.02)}$$

and

$$J(\zeta) = \int_0^\zeta [I(\xi) - I_\infty]\, d\xi, \qquad J(\zeta) \to J_\infty \quad \text{as } \zeta \to \infty. \qquad \text{(C.03)}$$

In these equations, the functions $F(\zeta)$ and $G(\zeta)$ are those discussed in Appendix A; the integrals I_∞ and J_∞ can be determined by numerical integration of the solutions once the "correct" values of $F'(0)$ and $G'(0)$ are known. Although it is possible to evaluate I_∞ and J_∞ by means of a quadrature, it is easier to write $y_6 = I$, $y_7 = J$ and add two first-order equations to the system of five equations given in equations (A.13) to (A.17). It can be shown that, correct to five significant figures,

$$I_\infty = 0.43992, \qquad J_\infty = -0.30374. \qquad \text{(C.04)}$$

Integration of equations (5.11) to (5.15), together with repeated use of the assumption (C.01), gives

$$\alpha'(0) \sim Pr\, H_\infty = -0.8845\, Pr \quad \text{as } Pr \to 0. \qquad \text{(C.05)}$$

$$\beta'(0) \sim \frac{4J_\infty}{H_\infty} = 1.3737 \quad \text{as } Pr \to 0. \qquad \text{(C.06)}$$

$$\gamma'(0) \sim -\frac{4}{Pr\, H_\infty} = \frac{4.5224}{Pr} \quad \text{as } Pr \to 0. \qquad \text{(C.07)}$$

$$\epsilon'(0) \sim 2\,Pr\,H_{\infty} = -1.7689\,Pr \quad \text{as } Pr \to 0. \tag{C.08}$$

$$\eta'(0) \sim I_{\infty}\,Pr = 0.43992\,Pr \quad \text{as } Pr \to 0. \tag{C.09}$$

Similar treatment of equation (5.44) for the function $\theta(\zeta)$ gives

$$\theta'(0) \sim \frac{2+n}{2}\,Pr\,H_{\infty} = -0.44223\,(2+n)\,Pr \quad \text{as } Pr \to 0. \tag{C.10}$$

C.3 Large values of the Prandtl number

In this case, the thermal boundary layer is very much thinner than the velocity one: the functions F, G and H in equations (3.11) to (3.15) can each be replaced by the first term in its series expansion in powers of ζ. Thus

$$F \sim a\zeta, \quad G \sim b\zeta, \quad H \sim -a\zeta^2, \tag{C.11}$$

where $a = F'(0) = 0.510233$ and $b = G'(0) = -0.615922$ (see equation (3.23)). (This asymptotic approximation does not work for the function $\beta(\zeta)$. This is not surprising since, for large ζ, the term in F^2 on the right-hand side of equation (5.12) approximates to a term in ζ^2: the behaviour of the term is, therefore, badly modelled for large values of ζ.)

It is convenient to define a new independent variable ξ where

$$\xi = (aPr)^{\frac{1}{3}}\zeta, \tag{C.12}$$

and functions $\alpha_0(\xi)$, $\gamma_0(\xi)$, $\epsilon_0(\zeta)$ and $\eta_0(\xi)$ such that

$$\alpha(\zeta) = \alpha_0(\xi), \tag{C.13}$$

$$\gamma(\zeta) = (aPr)^{-\frac{2}{3}}\gamma_0(\xi), \tag{C.14}$$

$$\epsilon(\zeta) = \epsilon_0(\xi), \tag{C.15}$$

$$\eta(\zeta) = (aPr)^{\frac{1}{3}}[\eta_0(\xi) - \frac{a^2+b^2}{2a}\xi^2]. \tag{C.16}$$

Using dots to denote differentiation with respect to ξ, the asymptotic forms of equations (5.11), (5.13), (5.14) and (5.15) become

$$\ddot{\alpha}_0 + \xi^2 \dot{\alpha}_0 = 0, \tag{C.17}$$

$$\ddot{\gamma}_0 + \xi^2 \dot{\gamma}_0 = -4\epsilon, \tag{C.18}$$

$$\ddot{\epsilon}_0 + \xi^2 \dot{\epsilon}_0 - 2\xi\epsilon_0 = 0, \tag{C.19}$$

$$\ddot{\eta}_0 + \xi^2 \dot{\eta}_0 - 2\xi\eta_0 = 0, \tag{C.20}$$

together with the boundary conditions

$$\alpha_0(0) = 1, \ \gamma_0(0) = 0, \ \epsilon_0(0) = 0, \ \eta_0(0) = 0 \tag{C.21}$$

and

$$\alpha_0, \gamma_0, \epsilon_0 \to 0, \quad \eta_0 \sim \frac{a^2 + b^2}{2a}\xi^2 \quad \text{as } \xi \to \infty. \tag{C.22}$$

These equations, which do not depend on Pr, can be solved in the same way as the full equations in Appendix B. Only functions $\alpha_2(\xi)$, $\gamma_1(\xi)$, $\gamma_2(\xi)$, $\epsilon_1(\xi)$ and $\epsilon_2(\xi)$ are required (corresponding to the functions $\alpha_2(\zeta)$, $\epsilon_1(\zeta)$, $\epsilon_2(\zeta)$, $\gamma_1(\zeta)$ and $\gamma_2(\zeta)$ of Appendix B). Equations (C.17) to (C.22) can be expressed as a set of 10 first-order equations with appropriate initial conditions.

It can be shown that, as $Pr \to \infty$,

$$\alpha'(0) \sim A_\alpha Pr^{\frac{1}{3}}, \quad A_\alpha = a^{\frac{1}{3}} B_\alpha = -0.62045, \tag{C.23}$$

$$\gamma'(0) \sim A_\gamma Pr^{-\frac{1}{3}}, \quad A_\gamma = a^{-\frac{1}{3}} B_\gamma = 2.0657, \tag{C.24}$$

$$\epsilon'(0) \sim A_\epsilon Pr^{\frac{1}{3}}, \quad A_\epsilon = a^{\frac{1}{3}} B_\epsilon = -0.84935, \tag{C.25}$$

$$\eta'(0) \sim A_\eta Pr^{\frac{2}{3}}, \quad A_\eta = a^{\frac{2}{3}} B_\eta = 0.58361, \tag{C.26}$$

where the numerical values have been computed on the VAX 8530. Integration as far as $\zeta = 4$ is sufficient to obtain values of A_α, A_γ and A_ϵ accurate to six significant figures; integration as far as $\zeta = 20$ is required to obtain A_η accurate to five significant figures.

C.4 Comparison with exact solution

Values of $\alpha'(0)$, $\beta'(0)$, $\gamma'(0)$, $\epsilon'(0)$ and $\eta'(0)$ have been computed using the full equations of Appendix B: they are shown in Tables C.1 and C.2. It is interesting to compare these values with the asymptotic solution discussed in this appendix.

Table C.1 Values of $\alpha'(0)$, $\beta'(0)$, $\gamma'(0)$, $\epsilon'(0)$ and $\eta'(0)$ for small Pr, computed using the full equations

Pr	10^{-9}	10^{-6}	10^{-3}	10^{-2}
$Pr^{-1}\,\alpha'(0)$	-0.8845	-0.8845	-0.8830	-0.8705
$\beta'(0)$	1.3736	1.3736	1.3696	1.3345
$Pr\,\gamma'(0)$	4.5225	4.5224	4.5080	4.3839
$Pr^{-1}\,\epsilon'(0)$	-1.7689	-1.7689	-1.7643	-1.7241
$Pr^{-1}\,\eta'(0)$	0.4399	0.4399	0.4395	0.4356

Table C.2 Values of $\alpha'(0)$, $\beta'(0)$, $\gamma'(0)$, $\epsilon'(0)$ and $\eta'(0)$ for large Pr, computed using the full equations

Pr	10^{3}	10^{6}	10^{9}
$Pr^{-1/3}\,\alpha'(0)$	-0.6016	-0.6186	-0.6203
$Pr^{-1/3}\,\beta'(0)$	2.1419	2.5690	2.6202
$Pr^{1/3}\,\gamma'(0)$	2.1370	2.0724	2.0663
$Pr^{-1/3}\,\epsilon'(0)$	-0.8297	-0.8474	-0.8492
$Pr^{-2/3}\,\eta'(0)$	0.5372	0.5785	0.5831

APPENDIX D

Solution of Energy-Integral Equation for Turbulent Flow

Equations (5.79) and (5.81) can be combined and written in the form

$$\frac{d}{dx}\left(\tilde{\chi} - \frac{\sigma}{1-\sigma}\tilde{H}x^{2\sigma+1}\right) = \frac{2K_w}{K_v}\left[\tilde{\chi}\tilde{H} + Ec_b^*\left(\frac{1}{\chi_{quad}} - \tfrac{1}{2}(1+R)\tilde{\chi}\right)x^2\right]x^{2\sigma}, \quad \text{(D.01)}$$

where $\tilde{\chi}$, $\tilde{H}$ are the ratios

$$\tilde{\chi} = \frac{\chi}{\chi_{quad}}, \quad \text{(D.02)}$$

$$\tilde{H} = \frac{H_0 - H_\infty}{(H_0 - H_\infty)_b} = \frac{1}{1+\frac{1}{2}Ec_b}\left[\frac{T_0 - T_\infty}{(T_0 - T_\infty)_b} + \tfrac{1}{2}Ec_b x^2\right], \quad \text{(D.03)}$$

and Ec_b^* is a modified Eckert number defined by

$$Ec_b^* = \frac{\Omega^2 b^2}{(H_0 - H_\infty)_b} = \frac{Ec_b}{1+\frac{1}{2}Ec_b}, \quad \text{(D.04)}$$

Ec_b being the Eckert number (see list of symbols) based on the temperature difference $(T_0 - T_\infty)_b$. The modified Eckert number, Ec_b^*, defined in equation (D.04) is analogous to the Nusselt numbers, Nu^* and Nu_{av}^*, discussed in Section 2.5.5: all are appropriate when dissipation effects are important, and reduce to the conventional values (Ec_b, Nu and Nu_{av} respectively) when dissipation is negligible. It is of interest to note that $Ec_b^* < 2$ as long as $(T_0 - T_\infty)_b > 0$ (that is $Ec_b > 0$).

To solve equation (D.01), it is convenient to write

$$y_1(x) = \tilde{\chi}^{-\frac{1}{1-\sigma}}\tilde{H}^{\frac{1}{\sigma}}x^{\frac{2\sigma+1}{\sigma}}; \tag{D.05}$$

equation (D.01) becomes

$$\frac{dy_1}{dx} = \frac{2K_w}{\sigma K_v}\left[\tilde{H}^{\frac{1}{\sigma}}x^{\frac{1+\sigma}{\sigma}} + Ec_b^*\left(\frac{1}{\chi_{quad}}y_1^{1-\sigma}x^{2(\sigma+1)} - \frac{1}{2}(1+R)\tilde{H}^{\frac{1-\sigma}{\sigma}}x^{\frac{1+3\sigma}{\sigma}}\right)\right],$$

$$y_1(0) = 0. \tag{D.06}$$

(The initial condition comes from equation (D.05) on the assumption that both $\tilde{\chi}$ and $\tilde{H}$ are finite when $x = 0$.)

Equation (D.06) is in a form suitable for numerical integration. As the integration proceeds, y_1 is found as a function of x and, using equations (D.02), (D.03) and (D.05),

$$\frac{\chi}{\chi_{quad}} = \tilde{H}^{\frac{1-\sigma}{\sigma}}x^{\frac{(2\sigma+1)(1-\sigma)}{\sigma}}y_1^{-(1-\sigma)}. \tag{D.07}$$

where $\tilde{H}$ is given by equation (D.03). Then, using equation (2.130) with $\chi = \chi'$ (which is valid for the free disc) and equation (5.53), it follows that

$$Nu^* = \frac{K_w\epsilon_m}{\pi}Re_\phi^\sigma Pr\chi_{quad}\tilde{H}^{\frac{1-\sigma}{\sigma}}x^{\frac{1+\sigma}{\sigma}}y_1^{-(1-\sigma)}. \tag{D.08}$$

Equation (2.120) with $\beta = 0$ gives

$$T_0 - T_{0,ad} = (T_0 - T_{0,\infty})_b\left[\frac{T_0 - T_\infty}{(T_0 - T_\infty)_b} - \frac{1}{2}Ec_b R x^2\right]; \tag{D.09}$$

hence, using equation (2.135), Nu_{av}^* can be computed by defining the functions $y_2(x)$, $y_3(x)$ such that

$$\frac{dy_2}{dx} = \left[\frac{T_0 - T_\infty}{(T_0 - T_\infty)_b} - \frac{1}{2}Ec_b R x^2\right]Nu^*, \qquad y_2(0) = 0, \tag{D.10}$$

$$\frac{dy_3}{dx} = \left[\frac{T_0 - T_\infty}{(T_0 - T_\infty)_b} - \frac{1}{2}Ec_b R x^2\right]x, \qquad y_3(0) = 0. \tag{D.11}$$

Equations (D.06), (D.10) and (D.11) form a set of three first order equations with specified initial conditions which can be solved by one of the standard methods. Then equation (2.133) gives

$$Nu^*_{av} = \frac{y_2(1)}{y_3(1)}. \tag{D.12}$$

For turbulent flow with a $\frac{1}{7}$-power-law velocity profile (see equation (5.73))

$$\sigma = \tfrac{4}{5},\ K_v = \tfrac{1}{6},\ K_w = \tfrac{23}{60},\ \epsilon_m = 0.2186, \tag{D.13}$$

and it is usual to take (see equations (5.74) and (5.82))

$$\chi_{quad} = Pr^{-\frac{2}{5}} = 1.147,\quad R = Pr^{\frac{1}{3}} = 0.892, \tag{D.14}$$

where the numerical values are for $Pr = 0.71$.

Equations (D.06), (D.10) and (D.11) then become

$$\frac{dy_1}{dx} = 5.75\left[\tilde{H}^{\frac{5}{4}}x^{\frac{9}{4}} + Ec_b^*\left(0.872\, y_1^{\frac{1}{5}}x^{\frac{18}{5}} - 0.946\,\tilde{H}^{\frac{1}{4}}x^{\frac{17}{4}}\right)\right],\qquad y_1(0) = 0, \tag{D.15}$$

$$\frac{dy_2}{dx} = \left[\frac{T_0 - T_\infty}{(T_0 - T_\infty)_b} - 0.446\, Ec_b x^2\right] Nu^*,\qquad y_2(0) = 0, \tag{D.16}$$

$$\frac{dy_3}{dx} = \left[\frac{T_0 - T_\infty}{(T_0 - T_\infty)_b} - 0.446\, Ec_b x^2\right] x,\qquad y_3(0) = 0, \tag{D.17}$$

with $\tilde{H}$ given by equation (D.03), Ec_b^* by equation (D.04) and, using equation (D.08),

$$\frac{Nu^*}{Re_\phi^{4/5}} = -0.0217\frac{1}{1 + \frac{1}{2}Ec_b}\left[\frac{T_0 - T_\infty}{(T_0 - T_\infty)_b} + \tfrac{1}{2}Ec_b x^2\right]^{\frac{1}{5}} x^{\frac{9}{4}}\, y_1^{-\frac{1}{5}} \tag{D.18}$$

The average Nusselt number Nu^*_{av} is given by equation (D.12).

List of References

Major citations in the text are on the pages <> indicated.

Abe, T, Kikuchi, J and Takeuchi, H 1979 An investigation of turbine disk cooling (experimental investigation and observation of hot gas flow into a wheel space). *13th CIMAC Cong., Vienna*, Paper No.GT-30 <226>

Batchelor, G K 1951 Note on a class of solutions of the Navier-Stokes equations representing steady rotationally-symmetric flow. *Quart.J.Mech.Appl.Math.*, **4**,29-41 <129>

Batchelor, G K 1967 *An introduction to fluid dynamics*. [Cambridge University Press, London] <13,151>

Bayley, F J and Conway, L 1964 Fluid friction and leakage between a stationary and rotating disc. *J.Mech.Engng Sci.*,**6**,164-172 <185>

Bayley, F J and Owen, J M 1969 Flow between a rotating and a stationary disc. *Aeronaut.Quart*.20,333-354 <179>

Bayley, F J and Owen, J M 1970 The fluid dynamics of a shrouded disk system with a radial outflow of coolant. *J.Engng for Power*,**92**,335-341 <210ff>

Benton, E R 1966 On the flow due to a rotating disk. *J.Fluid Mech.*,**24**,781-800 <44,247>

Blasius, H 1913 Das Ähnlichkeitsgesetz bei Reibungsvorgängen in Flüssigkeiten. *Forschg.Arb.Ing.-Wes. No*.134, *Berlin* <70,127>

Bödewadt, U T 1940 Die Drehströmung über festem Grunde. *Z.angew.Math.Mech.*,20,241-253 <59ff>

Brown, W B 1961 A stability criterion for three-dimensional laminar boundary layers. *Boundary layers and flow control*,**2**, [*ed*.Lachmann, G.V., Pergamon Press, London] <66>

Carper, Jr, H J, Saavedra, J J and Suwanprateep, T 1986 Liquid jet impingement cooling of a rotating disk. *J.Heat Transfer*,**108**,540-546 <203>

Cebeci, T and Abbott, D E 1975 Boundary layers on a rotating disc. *AIAA J.*,**13**,829-832 <80f>

Cebeci, T and Smith, A M O 1974 *Analysis of turbulent boundary layers*. [Academic Press, New York] <119,142>

Cebeci, T and Stewartson, K 1980 On stability and transition in three-dimensional flows. *AIAA J.*,**18**, 398-405 <67>

Cham, T-S and Head, M R 1969 Turbulent boundary-layer flow on a rotating disk. *J.Fluid Mech.*,**37**,129-147 <49f,81>

Chew, J W 1985 Effect of frictional heating and compressive work in rotating axisymmetric flow. *J.Heat Transfer*,**107**,984-986 <29>

Chew, J W 1989 A theoretical study of ingress for shrouded rotating disc systems with radial outflow. *To be presented at 34th ASME Int.Gas Turbine Conf., Toronto* <88,225>

Chew, J W and Rogers, R H 1988 An integral method for the calculation of turbulent forced convection in a rotating cavity with radial outflow. *Int.J.Heat Fluid Flow*,**9**,37-48 <116>

Chew, J W and Vaughan, C M 1988 Numerical predictions for the flow induced by an enclosed rotating disc. 33rd ASME Int.Gas Turbine Conf., Amsterdam, Paper No. 88-GT-127 <178>

Chin, D-T and Litt, M 1972a Mass transfer to point electrodes on the surface of a rotating disk. *J.Electrochem.Soc.*,**119**,1338-1343 <68>

Chin, D-T and Litt, M 1972b An electrochemical study of flow instability on a rotating disk. *J.Fluid Mech.*,**54**,613-625 <68>

Clarkson, M H, Chin, S C and Shacter, P 1980 Visualization of flow instabilities on a rotating disk. *AIAA J.*,**18**,1541-1543 <68>

Cobb E C and Saunders, O A 1956 Heat transfer from a rotating disk. *Proc.Roy.Soc.A*,**236**,343-351 <119f>

Cochran, W G 1934 The flow due to a rotating disk. *Proc.Camb.Phil.Soc.*,**30**,365-375 <44>

Cooper, P and Reshotko, E 1975 Turbulent flow between a rotating disk and a parallel wall. *AIAA J.*,**13**,573-578 <142>

Daily, J W and Nece, R E 1960 Chamber dimension effects on induced flow and frictional resistance of enclosed rotating disks. *J.Basic Engng*,**82**,217-232 <125ff>

Daily, J W, Ernst, W D and Asbedian V V 1964 Enclosed rotating discs with superimposed throughflow. *Dept.Civil Engng, Hydrodyn.Lab. MIT* Rep.No.64 <171,175,179>

Dibelius, G, Radtke, F and Ziemann, M 1984 Experiments on friction, velocity and pressure distribution of rotating discs. *Heat and mass transfer in rotating machinery*. [*ed*.Metzger, D E and Afgan, N H, Hemisphere, Washington] <185f>

Dijkstra, D and van Heijst, G J F 1983 The flow between two finite rotating disks enclosed by a cylinder. *J.Fluid Mech*.,**128**,123-154 <134f,137,140>

Dorfman, L A 1958 Resistance of a rotating rough disc. *Zh.tekh.Fiz*.,**28**,380-386 (Transl. *Sov.Phys.,Tech.Phys*., **3**,353-367) <91>

Dorfman, L A 1961 Effect of radial flow between the rotating disc and housing on their resistance and heat transfer. *Izv.Akad.Nauk-SSSR,OTN,Mekh.i Mash*.,*No*.**4**, 26-32 <162,164,190>

Dorfman, L A 1963 *Hydrodynamic resistance and the heat loss of rotating solids*. [Oliver and Boyd, Edinburgh] <13,29,78,109>

Ekman, V W 1905 On the influence of the earth's rotation on ocean-currents. *Ark.Mat.Astr.Fys*.,**2**,1-52 <20>

El-Oun, Z B 1986 Flow in a shrouded rotor-stator cavity with pre-swirl coolant. *D.Phil.thesis, University of Sussex* <238f,241>

El-Oun, Z B, Neller, P H and Turner, A B 1988 Sealing of a shrouded rotor-stator system with pre-swirl coolant. *J.Turbomachinery*,**110**,218-228 <238,242>

El-Oun, Z B and Owen, J M 1988 Pre-swirl blade-cooling effectiveness in an adiabatic rotor-stator system. 33*rd ASME Int.Gas Turbine Conf., Amsterdam*, Paper No. 88-GT-276 <204,208>

Erian, F F and Tong, Y H 1971 Turbulent flow due to a rotating disk. *Phys.Fluids*,**14**,2588-2591 <81f>

Escudier, M P 1984 Observations of the flow produced in a cylindrical container by a rotating endwall. *Experiments in Fluids*,2,189-196 <149ff>

Escudier, M P 1988 Vortex breakdown: observations and explanations. *Prog.Aerospace Sci*.,**25**,189-229 <149>

Faller, A J and Kaylor, R E 1966a A numerical study of the instability of the laminar Ekman boundary layer. *J.Atmos.Sci*.,**23**,466-480 <67>

Faller, A J and Kaylor, R E 1966b Investigations of stability and transition in rotating boundary layers. *Dynamics of fluids and plasmas* [*ed*. Pai, S L *et al*, Academic Press, New York] <68>

Federov, B I, Plavnik, G Z, Prokhorov, I V and Zhukhovitskii, L G 1976 Transitional flow conditions on a rotating disk. *J.Engng Phys*.,**31**,1448-1453 <68>

Goldstein, S 1935 On the resistance to the rotation of a disc immersed in a fluid. *Proc.Camb.Phil.Soc.*, **31**, 232-241 <70,74,77>

Gosman, A D, Koosinlin, M L. Lockwood, F C and Spalding, D B 1976 Transfer of heat in rotating systems. 21*st ASME Gas Turbine Conf.*, *New Orleans*, Paper No.76-GT-25 <185>

Gosman, A D, Lockwood, F C and Loughhead, J N 1976 Prediction of recirculating, swirling, turbulent flow in rotating disc systems. *J.Mech.Engng.Sci.*, 18, 142-148 <184,199>

Gosman, A D, Pun, W M, Runchal, A K, Spalding, D B and Wolfshtein, M 1969 *Heat and mass transfer in recirculating flows*. [Academic Press, London] <184>

Graber, D J, Daniels, W A and Johnson, B V 1987 Disk pumping test. *Air Force Wright Aeronaut.Lab.*, Report No.AFWAL-TR-87-2050 <185,236>

Granville P S 1973 The torque and turbulent boundary layer of rotating disks with smooth and rough surfaces, and in drag-reducing polymer solutions. *J.Ship Res.*, **17**, 181-195 <92>

Greenspan, H P 1968 *The theory of rotating fluids*. [Cambridge University Press, London] <13>

Gregory, N, Stuart, J T and Walker, W S 1955 On the stability of three-dimensional boundary layers with application to the flow due to a rotating disk. *Phil.Trans.Roy.Soc.A*, **248**, 155-199 <49f,65,81>

Gregory, N and Walker, W S 1960 Experiments on the effect of suction on the flow due to a rotating disk. *J.Fluid Mech.*, **9**, 225-234 <68>

Grohne, D 1955 Über die laminare Strömung in einer kreiszylindrischen Dose mit rotierendem Deckel. *Nachr. Akad.Wiss.Göttingen, Math.-Phys.Kl.*, 263-282 <130>

Haaser, F, Jack, J and McGreehan, W 1988 Windage rise and flowpath gas ingestion in turbine rim cavities. *J.Engng for Gas Turbines and Power*, **110**, 78-85 <236>

Haynes, C M and Owen, J M 1975 Heat transfer from a shrouded disk system with a radial outflow of coolant. *J.Engng for Power*, **97**, 28-36 <183f,195,197f>

Hennecke, D K, Sparrow, E M and Eckert, E R G 1971 Flow and heat transfer in a rotating enclosure with axial througflow. *Warme- und Stoffubertragung*, **4**, 222-235 <199>

Kanaev, A A 1953 Frictional power loss of a disc rotating in a liquid. *Zh.Tekh.Fiz.*, **23**, 317-322 <92>

Kapinos, V M 1957 Heat transfer from gas turbine discs with air cooling *Trudy Khark.Politekh.Inst.*, **24**, 111-133 <190>

Kapinos, V M 1965 Heat transfer from a disc rotating in a housing with a radial flow of coolant. *J.Engng.Phys.*, **8**, 35-38 <191>

Kármán, Th von 1921 Über laminare und turbulente Reibung. *Z.angew.Math.Mech.*, 1, 233-252 <5,43ff,70>

Kempf, G 1924 Über Reibungswiderstand rotierender Scheiben. *Vorträge aus dem Gebiete der Hydro- und Aerodynamik*, *Berlin* <50,82>

Ketola, H N and McGrew, J M 1968 Pressure, frictional resistance, and flow characteristics of the partially wetted rotating disk. *J.Lub.Tech.*,**90**,395-404 <145>

Kibel', I A 1947 Heating of a viscous liquid by a rotating disc. *Priklad.Mat.i Mekh.*,11,611-614 <94>

Kobayashi, R, Kohama, Y and Takamadate, Ch 1980 Spiral vortices in boundary layer transition regime on a rotating disk. *Acta Mech.*,**35**,71-82 <67f>

Kobayashi, N, Matsumato, M and Shizuya, M 1984 An experimental investigation of a gas turbine disk cooling system. *J.Engng for Gas Turbines and Power*,**106**,136-141 <226>

Köhler, M and Müller, U 1971 Theoretical and experimental investigations of the laminar flow between a stationary and a rotating disc. *Journal de Mécanique*,10,565-580 <161>

Koosinlin, M L, Launder, B E and Sharma, B I 1974 Prediction of momentum, heat and mass transfer in swirling turbulent boundary layers. *J.Heat Transfer*,**96**,204-209 <143>

Kreith, F, Doughman E and Kozlowski, H 1963 Mass and heat transfer from an enclosed rotating disk with and without source flow. *J.Heat Transfer*,**85**,153-163 <190>

Kreith, F, Taylor, J H and Chong, J P 1959 Heat and mass transfer from a rotating disk. *J.Heat Transfer*,,**81**, 95-105 <68,120>

Kreith, K and Viviand, H 1966 Heat transfer in laminar source flow between a stationary and a rotating disk. *Proc.Int.Heat Transfer Conf.*(3rd), *American Inst.Chem. Eng.*, *Chicago*,1,293-298 <190>

Kurokawa, J, Toyokura, T, Shinjo, M and Matsuo, K 1978 Roughness effects on the flow along an enclosed rotating disk. *Bull.JSME*,21,1725-1732 <153,155>

Lance, G N and Rogers, M H 1962 The axially symmetric flow of a viscous fluid between two infinite rotating disks. *Proc.Roy.Soc.A*,**266**,109-121 <130,133>

Lilly, D K 1966 On the instability of Ekman boundary flow. *J.Atmos.Sci.*,**23**,481-494 <67>

Lonsdale, G 1988 Solution of a rotating Navier-Stokes problem by a nonlinear multigrid algorithm. *J.Comp.Phys.*,**74**,177-190 <173>

Lugt, H J and Abboud, M 1987 Axisymmetric vortex breakdown with and without temperature effects in a container with a rotating lid. *J.Fluid Mech.*,**179**, 179-200 <151>

Mabuchi, I, Tanaka, T and Sakakibara, Y 1971 Studies on the convective heat transfer from a rotating disk (5th Report). *Bull.JSME*,**14**,581-589 <120>

McComas, S T and Hartnett, J P 1970 Temperature profiles and heat transfer associated with a single disk rotating in still air. *Proc.4th Heat Transfer Conf., Versailles*,3,Paper No.FC7.7 <120ff>

Malik, M R, Wilkinson, S P and Orszag, S A 1981 Instability and transition in rotating disk flow. *AIAA J.*,**19**,1131-1138 <67f>

Maroti, L A, Deak, G and Kreith F 1960 Flow phenomena of partially enclosed rotating disks. *J.Basic Engng*,**82**, 539-552 <147>

Meierhofer, B and Franklin, C J 1981 An investigation of a preswirled cooling airflow to a gas turbine disk by measuring the air temperature in the rotating channels. 26*th ASME Int.Gas Turbine Conf.,Houston*,Paper No. 81-GT-132 <204>

Mellor, G L, Chapple, P J and Stokes, V K 1968 On the flow between a rotating and a stationary disk. *J.Fluid Mech.*,**31**,95-112 <131)

Metzger, D E and Grochowsky, L D 1977 Heat transfer between an impinging jet and a rotating disk. *J.Heat Transfer*,**99**,663-667 <200>

Metzger, D E, Mathis, W J and Grochowsky, L D 1979 Jet cooling at the rim of a rotating disk. *J.Engng for Power*,**101**,68-78 <201ff>

Metzger, D E and Mitchell, J W 1966 Heat transfer from a shrouded rotating disk with film cooling. *J.Heat Transfer*,**88**,140-146 <200>

Millsaps, K and Pohlhausen, K 1952 Heat transfer by laminar flow from a rotating plate. *J.Aeronaut.Sci.*,**19**,120-126 <96,255>

Millward, J A and Robinson, P H 1989 Experimental investigation into the effects of rotating and static bolts on both windage heating and local heat transfer coefficients in a rotor/stator cavity. *To be presented at* 34*th ASME Int.Gas Turbine Conf., Toronto* <185>

Milne-Thomson, L M 1960 *Theoretical hydrodynamics* [Macmillan, London] <62>

Mitchell, J W and Metzger, D E 1965 Heat transfer from a shrouded rotating disk to a single fluid stream. *J.Heat Transfer*,**87**,485-492 <199>

Morse A P 1989 Assessment of laminar-turbulent transition in closed disc geometries. *To be presented at* 34*th ASME Int.Gas Turbine Conf., Toronto* <143,145>

Nece, R E and Daily, J W 1960 Roughness effects on frictional rsistance of enclosed rotating disks. *J.Basic Engng*,**82**,553-562 <152f>

Nikitenko, N I 1963 Experimental investigation of heat exchange of a disk and a screen. *J.Engng Phys.*,**6**,1-11 <121,188>

Nikuradse, J 1932 Gesetzmässigkeit der turbulenten Strömung in glatten Rohren. *Forschg.Arb.Ing.-Wes.* No.356 <69>

Northrop, A and Owen, J M 1988 Heat transfer measurements in rotating-disc systems. Part 1: The free disc. *Int.J.Heat and Fluid Flow*,**9**,19-26 <123f>

Okaya, T and Hasegawa, M 1939 On the friction to the disc rotating in a cylinder. *Japanese J.Phys.*,**13**, 29-38 <137>

Ong, C L 1988 Computation of fluid flow and heat transfer in rotating disc systems. *D.Phil.thesis, University of Sussex* <81,114,119f,124>

Owen, J M 1969 Flow between a rotating and a stationary disc. *D.Phil.thesis, University of Sussex* <82f>

Owen, J M 1971 The Reynolds analogy applied to flow between a rotating and a stationary disc. *Int.J.Heat Mass Transfer*,**14**,451-460 <29>

Owen J M 1988 An approximate solution for the flow between a rotating and a stationary disc. 33*rd ASME Int.Gas Turbine Conf., Amsterdam*, Paper No.88-GT-293 <136,166,170>

Owen, J M and Haynes, C M 1976 Design formulae for the heat loss and frictional resistance of air-cooled rotating discs. *Improvements in Fluid Mechanics and Systems for Energy Conversion*,**IV**,127-160 [Hoepli, Milan] <181f,194,196>

Owen, J M, Haynes, C M and Bayley, F J 1974 Heat transfer from an air-cooled rotating disk. *Proc.Roy. Soc.A*,**336**,453-473 <120,122,179ff,192ff>

Owen, J M and Phadke, U P 1980 An investigation of ingress for a simple shrouded rotating disk system with a radial outflow of coolant. 25*th ASME Int.Gas Turbine Conf., New Orleans*, Paper No.80-GT-49 <215>

Owen, J M and Pincombe, J R 1979 Vortex breakdown in a rotating cylindrical cavity. *J.Fluid Mech.*,**90**,109-127 <199>

Owen, J M, Pincombe, J R and Rogers, R H 1985 Source-sink flow inside a rotating cylindrical cavity. *J.Fluid Mech.*,**155**,233-265 <26,165>

Pantell, K 1950 Versuche über Scheibenreibung. *Forschung auf dem Gebiete des Ingenieurwesens*, **16**, 97-108 <139>

Pearson, C E 1965 Numerical solutions for the time-dependent viscous flow between two rotating coaxial disks. *J.Fluid Mech.*,**21**,623-633 <131>

Phadke, U P 1982 Aerodynamic aspects of the sealing of gas turbine rotor-stator systems. *D.Phil.thesis, University of Sussex* <215,217,225>

Phadke, U P and Owen, J M 1983 An investigation of ingress for an air-cooled shrouded rotating disc system with radial clearance seals. *J.Engng.Power*, **105**, 178-183 <215>

Phadke, U P and Owen, J M 1988a Aerodynamic aspects of the sealing of gas-turbine rotor-stator systems. Part 1: The behavior of simple shrouded rotating-disk systems in a quiescent environment. *Int.J.Heat and Fluid Flow*, **9**, 98-105 <215ff>

Phadke, U P and Owen, J M 1988b Aerodynamic aspects of the sealing of gas-turbine rotor-stator systems. Part 2: The performance of simple seals in a quasi-axisymmetric external flow. *Int.J.Heat and Fluid Flow*, **9**, 106-112 <226ff>

Phadke, U P and Owen, J M 1988c Aerodynamic aspects of the sealing of gas-turbine rotor-stator systems. Part 3: The effect of nonaxisymmetric external flow on seal performance. *Int.J.Heat and Fluid Flow*, **9**, 113-117 <226,232ff,238>

Picha, K G and Eckert, E R G 1958 Study of the air flow between coaxial discs rotating with arbitrary velocities in an open or enclosed space. *Proc.3rd U.S.Nat.Cong. Appl.Mech.*, 791-798 <130>

Popiel, Cz O and Boguslawski L 1975 Local heat-transfer coefficients on the rotating disk in still air. *Int.J. Heat Mass Transfer*, **18**, 167-170 <120>

Prandtl, L 1921 Über den Reibungswiderstand strömender Luft. *Ergebnisse AVA Göttingen, 1st Series*, 136 <70>

Reshotko, E and Rosenthal, R L 1968 Fluid dynamic considerations in the design of slinger seals. *J.American Soc.Lub.Eng*, **24**, 303-314 <141f>

Richardson, P D and Saunders, O A 1963 Studies of flow and heat transfer associated with a rotating disc. *J.Mech.Engng.Sci.*, **5**, 336-342 <120,148>

Rogers, M H and Lance, G N 1960 The rotationally symmetric flow of a viscous fluid in the presence of an infinite rotating disk. *J.Fluid Mech.*, **7**, 617-631 <51,54,59,244>

Rogers, R H 1989 Heat transfer from a rotating disc in a rotating fluid. *Report No.89/TFMRC/117, School of Engineering and Applied Sciences, University of Sussex* <108,113>

Rott, N and Lewellen, W S 1966 Boundary layers and their interactions in rotating flows. *Prog. Aeronaut.Sci.*, **7**, 111-144 <87>

Sambo, A S 1983 A theoretical and experimental study of the flow between a rotating and a stationary disc. *D.Phil.thesis, University of Sussex* <138,141>

Schlichting, H 1979 *Boundary-layer theory*. [McGraw Hill, New York] <*passim*>

Schmidt, W 1921 Ein einfaches Messverfahren für Drehmomente. *Zeitschrift des Vereines deutscher Ingenieure*,**65**,441-444 <82>

Schultz-Grunow, F 1935 Der Reibungswiderstand rotierender Scheiben in Gehäusen. *Z.angew.Math.Mech.*,**15**, 191-204 <138>

Sedach, V S 1957 Kinematics of the air flow cooling a gas turbine disc. *Trudy Khark.Politekh.Inst.*,**24**, 69-87 <139>

Shampine, L F and Gordon, M K 1975 *Computer solution of ordinary differential equations: the initial value problem* [W H Freeman, San Francisco] <59,247>

Shchukin, V K and Olimpiev, V V 1975 Heat transfer of disc rotating in a housing with transitional and turbulent boundary layers. *Soviet Aeronautics*,**18**, 77-81 <189>

Smith, N H 1947 Exploratory investigation of laminar-boundary-layer oscillations on a rotating disk. *NACA Tech.Note*, No.1227 <65>

Socio, L M de, Sparrow, E M and Eckert, E R G 1976 Analysis of rotating, recirculating turbulent flow and heat transfer in an enclosure with fluid throughflow. *Int.J.Heat Mass Transfer*,**19**,345-347 <199>

Soo, S L 1958 Laminar flow over an enclosed rotating disk. *Trans.ASME*,**80**,287-296 <127,158,160f>

Soo, S L, Besant, R W and Sarafa, Z N 1962 The nature of heat transfer from an enclosed rotating disk. *Z.angew.Math.Phys.*,**13**,297-309 <190>

Spalding, D B and Patankar, S V 1967 *Heat and mass transfer in boundary layers* [Morgan-Grampian, London] <180,193>

Sparrow, E M, Buszkiewicz, T C and Eckert, E R G 1975 Heat transfer and temperature field experiments in a cavity with rotation, recirculation and coolant throughflow. *J.Heat Transfer*,**97**,22-28 <199>

Sparrow, E M and Goldstein, Jr, L 1976 Effect of rotation and coolant throughflow on the heat transfer and temperature field in an enclosure. *J.Heat Transfer*,**98**,387-394 <199>

Sparrow, E M and Gregg, J L 1959 Heat transfer from a rotating disk to fluids of any Prandtl number. *J.Heat Transfer*,**81**,249-251 <96>

Sparrow, E M, Shamsundar, N and Eckert, E R G 1973 Heat transfer in rotating cylindrical enclosures with axial inflow and outflow of coolant. *J.Engng for Power*,**95**, 278-280 <199>

Stewartson, K 1953 On the flow between two rotating coaxial discs. *Proc.Camb.Phil.Soc.*,**49**,333-341 <130>

Szeri, A Z and Adams, M L 1978 Laminar throughflow between closely spaced rotating disks. *J.Fluid Mech.*, **86**,1-14 <161>

Szeri, A Z, Schneider, S J, Labbe, F and Kaufman, H N 1983 Flow between rotating disks. Part I. Basic flow. *J.Fluid Mech.*,**134**,103-131 <141,162>

Targ, S M 1951 Basic problems in the theory of laminar flow. *Moscow: Gostekhizdat* <48>

Theodorsen, T and Regier, A 1944 Experiments on drag of revolving disks, cylinders, and streamline rods at high speeds. *NACA Report*, No.793 <51,67,82f>

Tifford, A N 1951 Note on "Heat transfer by laminar flow from a rotating plate". *J.Aeronaut.Sci.*,18,567-568 <102>

Tifford, A N and Chu, S T 1954 On the flow and temperature fields in forced flow against a rotating disc. *Proc.2nd U.S.Nat.Cong.Appl.Mech.* [publ.by ASME, 1955],793-800 <63f>

Tobak, M 1973 On local inflexional instability in boundary-layer flows. *J.Appl.Math.Phys.*,**24**,330-354 <67>

Truckenbrodt, E 1954 Die turbulente Strömung an einer angeblasenen rotierenden Scheibe. *Z.angew.Math.Mech.*, **34**,150-162 <89>

Vaughan, C 1986 A numerical investigation into the effect of an external flow field on the sealing of a rotor-stator cavity. *D.Phil.thesis, University of Sussex* <137f,143f,167f,171,173,176,178f,229f>

Wieghardt, K 1946 Turbulente Grenzschichten. *Göttinger Monographie*, Part **B5** <70>

Wilkinson, S P and Malik, M R 1985 Stability experiments in the flow over a rotating disk. *AIAA J.*,**23**,588-595 <68>

Yamashita, I and Takematsu, M 1974 A numerical study on the instability of three-dimensional boundary layers. *Rep.Res.Inst.Appl.Mech.*,**22**,1-14 <67>

Yu, J P, Sparrow, E M and Eckert, E R G 1973 Experiments on a shrouded, parallel disk system with rotation and coolant throughflow. *Int.J.Heat Mass Transfer*,**16**, 311-328 <197>

Zandbergen, P J and Dijkstra, D 1987 Von Kármán swirling flows. *Ann.Rev.Fluid Mech.*,**19**,465-491 <132,244>

Zimmermann, H, Firsching, A, Dibelius, G H and Ziemann, M 1986 Friction losses and flow distribution for rotating disks with shielded and protruding bolts. *J.Engng for Gas Turbines and Power*,**108**,547-561 <185>

Index